FIGURES FOR FUN

Stories, Puzzles and Conundrums

Yakov Perelman

Dover Publications, Inc.
Mineola, New York

Bibliographical Note

Figures for Fun: Stories, Puzzles and Conundrums, first published by Dover Publications, Inc., in 2015, is an unabridged republication of the American edition of *Figures for Fun: Stories and Conundrums*, originally published by the Frederick Ungar Publishing Company, New York, in 1965.

Library of Congress Cataloging-in-Publication Data

Perelman, IA. I. (IAkov Isidorovich), 1882-1942. [Zhivaia matematika. English]
Figures for fun : stories, puzzles, and conundrums / Yakov Perelman.
pages cm
Originally published in English by: Moscow : Foreign Languages Pub. House, 1957.
ISBN-13: 978-0-486-79568-3
ISBN-10: 0-486-79568-3
1. Mathematical recreations. I. Title.

QA95.P363 2015
510—dc23

2014040410

International Standard Book Number

ISBN-13: 978-0-486-79568-3
ISBN-10: 0-486-79568-3

Manufactured in the United States by Courier Corporation
79568301 2015
www.doverpublications.com

CONTENTS

CHAPTER III

Another Dozen Puzzlers

CHAPTER IV

Counting

CHAPTER V

Bafflers With Numbers

CHAPTER VI

Number Giants

CHAPTER VII

Without Instruments of Measurement

CHAPTER VIII

Geometric Brain-Teasers

CHAPTER IX

The Geometry of Rain and Snow

CHAPTER X

Mathematics and the Deluge

CHAPTER XI

Thirty Different Problems

PREFACE

To read and enjoy this book it will suffice to possess a modest knowledge of mathematics, i.e., knowledge of arithmetical rules and elementary geometry. Very few problems require the ability of forming and solving equations—and the simplest at that.

The table of contents, as you may see, is quite diversified: the subjects range from a motley collection of conundrums and mathematical stunts to useful practical problems on counting and measuring. The author has done everything to make his book as fresh as possible, avoiding repetition of all that has already appeared in his other works (*Tricks and Amusements*, *Interesting Problems*, etc.). The reader will find a hundred or so brain-teasers that have not been included in earlier books. Chapter VI—"Number Giants"—is adapted from one of the author's earlier pamphlets, with four new stories added.

CHAPTER I

BRAIN-TEASERS FOR LUNCH

It was raining.... We had just sat down for lunch at our holiday home when one of the guests asked us whether we would like to hear what had happened to him in the morning.

Everyone assented, and he began.

1. A SQUIRREL IN THE GLADE.—"I had quite a bit of fun playing hide-and-seek with a squirrel," he said. "You know that little round glade with a lone birch in the centre? It was on this tree that a squirrel was hiding from me. As I emerged from a thicket, I saw its snout and two bright little eyes peeping from behind the trunk. I wanted to see the little animal, so I started circling round along the edge of the glade, mindful of keeping the distance in order not to scare it. I did four rounds, but the little cheat kept backing away from me, eyeing me suspiciously from behind the tree. Try as I did, I just could not see its back."

"But you have just said yourself that you circled round the tree four times," one of the listeners interjected.

"Round the tree, yes, but not round the squirrel."

"But the squirrel was on the tree, wasn't it?"

"So it was."

"Well, that means you circled round the squirrel too."

"Call that circling round the squirrel when I didn't see its back?"

"What has its back to do with the whole thing? The squirrel was on the tree in the centre of the glade and you circled round the tree. In other words, you circled round the squirrel."

"Oh no, I didn't. Let us assume that I'm circling round you and you keep turning, showing me just your face. Call that circling round you?"

"Of course, what else can you call it?"

"You mean I'm circling round you though I'm never behind you and never see your back?"

"Forget the back! You're circling round me and that's what counts. What has the back to do with it?"

"Wait. Tell me, what's circling round anything? The way I understand it, it's moving in such a manner so as to see the object I'm moving around from all sides. Am I right, professor?" He turned to an old man at our table.

"Your whole argument is essentially one about a word," the professor replied. "What you should do first is agree on the definition of 'circling.' How do you understand the words 'circle round an object'? There are two ways of understanding that. Firstly, it's moving round an object that is in the centre of a circle. Secondly, it's moving round an object in such a way as to see all its sides. If you insist on the first meaning, then you walked round the squirrel four times. If it's the second that you hold to, then you did not walk round it at all. There's really no ground for an argument here, that is, if you two speak the same language and understand words in the same way."

"All right, I grant there are two meanings. But which is the correct one?"

"That's not the way to put the question. You can agree about anything. The question is, which of the two meanings is the more generally accepted? In my opinion, it's the first and here's why. The sun, as you know, does a complete circuit in 26 days. . . ."

"Does the sun revolve?"

"Of course, it does, like the earth. Just imagine, for instance, that it would take not 26 days, but $365^1/_4$ days, i.e., a whole year, to do so. If this were the case, the earth would see only one side of the sun, that is, only its 'face.' And yet, can anyone claim that the earth does not revolve round the sun?"

"Yes, now it's clear that I circled round the squirrel after all."

"I've a suggestion, comrades!" one of the company shouted. "It's raining now, no one is going out, so let's play riddles. The squirrel riddle was a good beginning. Let each think of some brain-teaser."

"I give up if they have anything to do with algebra or geometry," a young woman said.

"Me too," another joined in.

"No, we must all play, but we'll promise to refrain from any algebraical or geometrical formulas, except, perhaps, the most elementary ones. Any objections?"

"None!" the others chorussed. "Let's go."

"One more thing. Let professor be our judge."

2. SCHOOL-GROUPS.—"We have five extra-curricular groups at school," a Young Pioneer began. "They're political, literary, photographic, chess and choral groups. The political group meets every other day, the literary every third day, the photographic every fourth day, the chess every fifth day and the choral every sixth day. These five groups first met on January 1 and thenceforth meetings were held according to schedule. The question is, how many times did all the five meet on the one and same day in the first quarter (January 1 excluded)?"

"Was it a Leap Year?"

"No."

"In other words, there were 90 days in the first quarter."

"Right."

"Let me add another question," the professor broke in. "It's this: how many days were there when none of the groups met in that first quarter?"

"So, there's a catch to it? There'll be no other evening when all the five groups meet and no evening when some do not meet. That's clear!"

"Why?"

"Don't know. But I've a feeling there's a catch."

"Comrades!" said the man who had suggested the game. "We won't reveal the results now. Let's have more time to think about them. Professor will announce the answers at supper."

3. WHO COUNTED MORE?—"Two persons, one standing at the door of his house and the other walking up and down the pavement, were counting passers-by for a whole hour. Who counted more?"

"Naturally the one walking up and down," said somebody at the end of the table.

"We'll know the answer at supper," the professor said. "Next!"

4. GRANDFATHER AND GRANDSON.—"In 1932 I was as old as the last two digits of my birth year. When I mentioned this interesting coincidence to my grandfather, he surprised me by saying that the same applied to him too. I thought that impossible. . . ."

"Of course that's impossible," a young woman said.

"Believe me, it's quite possible and grandfather proved it too. How old was each of us in 1932?"

5. RAILWAY TICKETS.—"I'm a railway ticket seller," said the next person, a young lady. "People think this job is easy. They probably have no idea how many tickets one has to sell, even at a small station. There are 25 stations on my line and different tickets for each section up and down the line. How many different kinds of tickets do you think I have at my station?"

"Your turn next," the professor said to a flier.

6. A DIRIGIBLE'S FLIGHT.—"A dirigible took off from Leningrad in a northerly direction. Five hundred kilometres away it turned and flew 500 kilometres eastward. After that it turned south and covered another 500 kilometres. Then it flew 500 kilometres in a westerly direction, and landed. The question is, where did it land: west, east, north or south of Leningrad?"

"That's an easy one," someone said. "Five hundred steps forward, 500 to the right, 500 back and 500 to the left, and you're naturally back where you'd started from!"

"Easy? Well then, where did the dirigible land?"

"In Leningrad, of course. Where else?"

"Wrong!"

"Then I don't understand."

"Yes, there's some catch to this puzzle," another joined in. "Didn't the dirigible land in Leningrad?"

"Won't you repeat your problem?"

The flier did. The listeners looked at each other.

"All right," the professor said. "We have enough time to think about the answer. Let's have the next one now."

7. SHADOW.—"My puzzle," said the next man, "is also about a dirigible. What's longer, the dirigible or its perfect shadow?"

"Is that all?"

Sunrays spread from behind a cloud.

"It is."

"Well, then. The shadow is naturally longer than the dirigible: sunrays spread fanlike, don't they?"

"I wouldn't say so," another interjected. "Sunrays are parallel to each other and that being so, the dirigible and its shadow are of the same size."

"No, they aren't. Have you ever seen rays spreading from behind a cloud? If you have, you've probably noticed how much they spread. The shadow of the dirigible must be considerably bigger than the dirigible itself, just as the shadow of a cloud is bigger than the cloud itself."

"Then why is it that people say that sunrays are parallel to each other? Seamen, astronomers, for instance."

The professor put a stop to the argument by asking the next person to go ahead with his conundrum.

8. MATCHES.—The man emptied a box of matches on the table and divided them into three heaps.

"You aren't going to start a bonfire, are you?" someone quipped.

"No, they're for my brain-teaser. Here you are—three uneven heaps. There are altogether 48 matches. I won't tell how many there are in each heap. Look well. If I take as many matches from the first heap as there are in the second and add them to the second, and then take as many from the second as there are in the third, and add them to the third, and finally take as many from the third as there are in the first and add them to the first—well, if I do all this, the heaps will all have the same number of matches. How many were there originally in each heap?"

9. THE "WONDERFUL" STUMP.—"My puzzle is the one I was once asked by a village mathematician to solve," the next person began. "It was really a story, and quite humorous at that. One day, a peasant met an old man in a forest. The two fell into a conversation and the latter said:

" 'There's a wonderful little stump in this forest. It helps people in need.'

" 'It does? What does it do, cure people?'

" 'Not exactly. It doubles one's money. Put your pouch among the roots, count one hundred and—presto!—the money's doubled. It's a wonderful stump, that!'

" 'Can I try it?' the peasant asked excitedly.

" 'Why not? Only you must pay.'

" 'Pay whom and how much?'

" 'The man who shows you the stump. That's me. As to how much, that's another matter.'

"The two men began to bargain. When the old man learned that the peasant did not have much money, he agreed to take 1 ruble 20 kopeks every time the money doubled.

"The two went deep into the forest where, after a long search, the old man brought the peasant to a moss-covered fir stump in bushes. He then took the peasant's pouch and shoved it among the roots. After that they counted one hundred. The old man took a long time to find the pouch and returned it to the peasant.

"The latter opened the pouch and, lo! The money really had doubled! He counted off the ruble and 20 kopeks, as agreed upon, and asked the old man to repeat the whole thing.

"Once again they counted one hundred, once again the old man began his search for the pouch and once again there was a miracle—the money had doubled again. And just as they had agreed, the old man got his ruble and 20 kopeks.

"Then they hid the pouch for the third time and this time too the money doubled. But after the peasant had paid the old man his ruble and 20 kopeks, there was nothing left in the pouch. The poor fellow had lost all his money in the process. There was no more money to be doubled and he walked off crest-fallen.

"The secret, of course, is clear to all—it was not for nothing that the old man took so long to find the pouch. But there is another question I would like to ask you: how much did the peasant have originally?"

10. THE DECEMBER PUZZLE.—"Well, comrades," began the next man. "I'm a linguist, not a mathematician, so you needn't expect a mathematical problem. I'll ask you one of another kind, one close to my sphere of activity. It's about the calendar."

"Go ahead."

"December is the twelfth month of the year. Do you know what the name really means? The word comes from the Greek 'deka'—ten. Hence, decalitre which means ten litres, decade—ten years, etc. December, to all appearances, should be the tenth month and yet it isn't. How d'you explain that?"

11. AN ARITHMETICAL TRICK.—"I'll give you an arithmetical trick and ask you to explain it. One of you—you, professor, if you like—write down a three-digit number, but don't tell me what it is."

"Can we have any noughts in it?"

"I set no reservations. You can write down any three numerals you want."

"All right, I've done it. What next?"

"Write the same number alongside. Now you have a six-digit number."

"Right."

"Pass the slip to your neighbour, the one farther away from me, and let him divide this six-digit number by seven."

"It's easy for you to say that, and what if it can't be done?"

"Don't worry, it can."

"How can you be so sure when you haven't seen the number?"

"We'll talk after you've divided it."

"You're right. It does divide."

"Now pass the result to your neighbour, but don't tell me what it is. Let him divide it by 11."

"Think you'll have your own way again?"

"Go ahead, divide it. There'll be no remainder."

"You're right again. Now what?"

"Pass the result on and let the next man divide it, say, by 13."

"That's a bad choice. There are very few numbers that are divisible by 13. . . . You're certainly lucky, this one is!"

"Now give me the slip, but fold it so that I don't see the number."

Without unfolding the slip, the man passed it on to the professor.

"Here's your number. Correct?"

"Absolutely." The professor was surprised. "That *is* the number I wrote down. . . . Well, everyone has had his turn, the rain has stopped, so let's go out. We'll know the answers tonight. You may give me all the slips now."

Answers 1 to 11

1. The squirrel puzzle was explained earlier, so we'll pass on to the next.

2. We can easily answer the first question: how many times did all the five groups meet on one and the same day in the first quarter (January 1 excluded) by finding the least common multiple of 2, 3, 4, 5 and 6. That isn't difficult. It's 60. Therefore, the five will all meet again on the 61st day—the political group after 30 two-day intervals, the literary after 20 three-day intervals, the photographic after 15 four-day intervals, the chess after 12 five-day intervals and the choral after 10 six-day intervals. In other words, they can meet on the one and same day only once in 60 days. And since there are 90 days in the first

quarter, it means there can be only one other day on which they all meet.

It is much more difficult to find the answer to the second question: how many days are there when none of the groups meets in the first quarter? To find that, it is necessary to write down all the numbers from 1 to 90 and then cross out all the days when the political group meets: e.g., 1, 3, 5, 7, 9, etc. After that one must cross out the literary group days: e.g., 4, 7, 10, etc. When all the photographic, chess and choral groups' days have also been crossed out, the numbers that remain are the days when there is no group meeting.

Do that and you'll see that there are 24 such days—eight in January, i.e., 2, 8, 12, 14, 18, 20, 24 and 30, seven in February and nine in March.

3. Both of them counted the same number of passers-by. While the one who stood at the door counted all those who passed both ways, the one who was walking counted all the people he met going up and down the pavement.

4. At first it may seem that the problem is incorrectly worded, that both grandfather and grandson are of the same age. We shall soon see that there is nothing wrong with the problem.

It is obvious that the grandson was born in the 20th century. Therefore, the first two digits of his birth year are 19 (the number of hundreds). The other two digits added to themselves equal 32. The number therefore is 16: the grandson was born in 1916 and in 1932 he was 16.

The grandfather, naturally, was born in the 19th century. Therefore, the first two digits of his birth year are 18. The remaining digits multiplied by 2 must equal 132. The number sought is half of 132, i.e., 66. The grandfather was born in 1866 and in 1932 he was 66.

Thus, in 1932 the grandson and the grandfather were each as old as the last two digits of their birth years.

5. At each of the 25 stations passengers can get tickets for any of the other 24. Therefore, the number of different tickets required is: $25 \times 24 = 600$.

6. There is nothing contradictory in this problem. The dirigible did not fly along the contours of a square. It should be borne in mind that the earth is round and that the meridians converge at the poles (fig. 2). Therefore, flying 500 kilometres along the parallel 500 kilometres north of Leningrad latitude, the dirigible covered *more degrees* going eastward than it did when it was returning along Leningrad latitude. As a result, the dirigible completed its flight *east* of Leningrad.

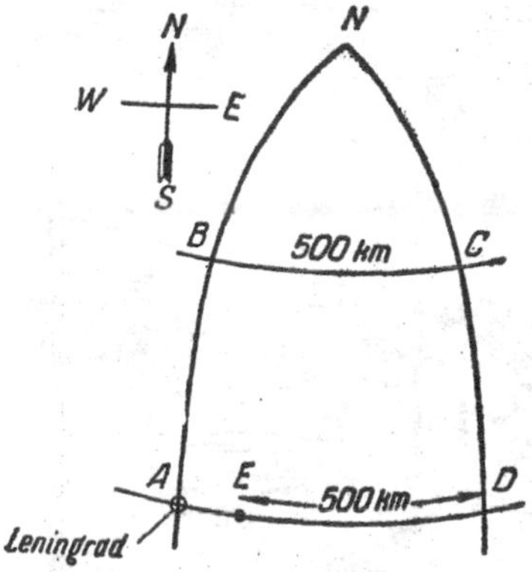

Fig. 2.

How many kilometres away? That can be calculated. Fig. 2 shows the route taken by the dirigible: *ABCDE*. *N* is the North Pole where meridians *AB* and *DC* meet. The dirigible first flew 500 kilometres northward, i.e., along meridian *AN*. Since the degree of a meridian is 111 kilometres long, the 500-kilometre-long arc of the meridian is equal to 500:111=4°5′. Leningrad lies on the 60th parallel. *B*, therefore, is on 60°+4°5′=64°5′. The airship then flew eastward, i.e., along the *BC* parallel, covering 500 kilometres. The length of one degree of this parallel may be calculated (or learned from tables); it is equal to 48 kilometres. Therefore, it is easy to determine how many degrees the dirigible covered in its eastward flight: 500:48=10°4′. Continuing, the airship flew southward, i.e., along meridian *CD*, and, having covered 500 kilometres, returned to the Leningrad parallel. Thence the way lay westward, i.e., along *DA*; the 500 kilometres of this way are obviously less than the distance between *A* and *D*. There are as many degrees in *AD* as in *BC*, i.e., 10°4′. But the length of 1° at the 60th parallel equals 55.5 kilometres. Therefore, the distance between *A* and *D* is equal to 55.5×10.4=577 kilometres. We see thus that the dirigible could not have very well landed in Leningrad: it landed 77 kilometres away, on Lake Ladoga.

7. In discussing this problem, the people in our story committed several mistakes. It is wrong to say that the rays of the sun spread fanlike. The earth is so small in comparison to its distance from the sun that the sunrays falling on any part of its surface radiate at an almost absolutely imperceptible angle; in fact, rays may be said to be parallel to each other. We occasionally see them spreading fanlike (for instance, when the sun is behind a cloud, fig. 1). This, however, is nothing but a case of perspective.

Parallel lines, as they recede from the station point, always appear to the eye to meet far away in a point, e.g., railway tracks or a long avenue (see fig. 3).

But the fact that sunrays fall to the ground in parallel beams does not mean that the perfect shadow of a dirigible is as long as the dirigible itself. Fig. 4 shows that the perfect shadow of the dirigible narrows down in space on the way to the surface of the earth and that, con-

Fig. 3.

sequently, the shadow the dirigible casts should be shorter than the dirigible: *CD* is shorter than *AB*.

It is quite possible to compute the difference, provided, of course, we know at what altitude the dirigible is flying. Let us assume that the altitude is 1,000 metres. The angle formed by lines *AC* and *BD* is equal to the angle from which the sun is seen from the earth. We know that this angle is equal to $^1/_2°$. On the other hand, we know that the distance between the eye and any object seen from an angle of $^1/_2°$ is equal to the length of 115 diameters of this object. Hence, section *MN* (the section seen from the surface of the earth at an angle of $^1/_2°$) should be

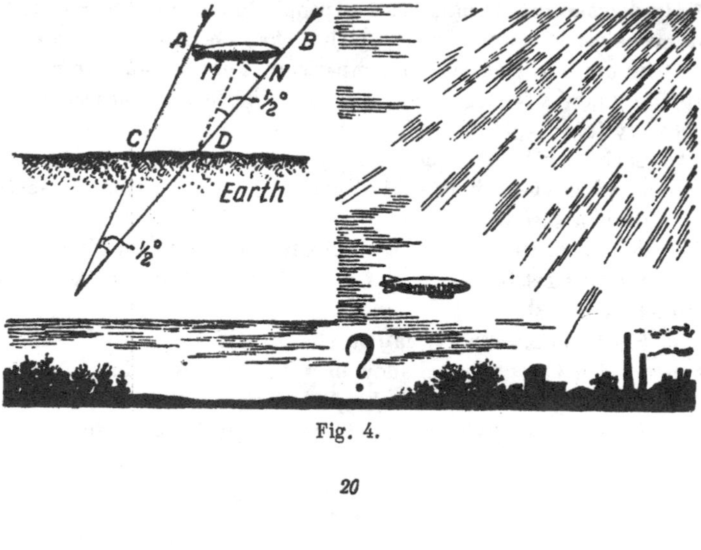

Fig. 4.

1/115 part of AC. Line AC is longer than the perpendicular distance between point A and the surface of the earth. If the angle formed by sunrays and the surface of the earth is equal to 45°, then AC (given the altitude of the dirigible at 1,000 metres) is approximately 1,400 metres long and section MN is consequently equal to 1,400:115=12 metres.

But then the difference between the dirigible and its shadow, i.e., section MB, is bigger than MN (1.4 times, to be exact), because angle MBD is almost equal to 45°. Therefore, MB is equal to 12×1.4, and that gives us almost 17 metres.

All this applies to the *perfect*—black and sharp—shadow of the dirigible, and not to *penumbra*, which is weak and hazy.

Incidentally, our computation shows that if instead of a dirigible we had a balloon almost 17 metres in diameter, there would be no *perfect* shadow. All we would see would be a hazy penumbra.

8. This problem is solved from the end. Let us proceed from the fact that, after all the transpositions, the number of matches in each heap is the same. Since the total number of matches (48) has not changed in the process, it follows that there were 16 in each heap.

And so, what we have in the end is:

First Heap	Second Heap	Third Heap
16	16	16

Immediately before that we had added to the first heap as many matches as there were in it, i.e., we had *doubled* the number. Thus, before that final transposition, there were only 8 matches in the first heap. In the third heap, from which we took these 8 matches, there were:

$$16+8=24.$$

Now we have the following numbers:

First Heap	Second Heap	Third Heap
8	16	24

Further, we know that from the second heap we took as many matches as there were in the third heap. That means 24 was *double* the original number. This shows us how many matches we had in each heap after the first transposition:

First Heap	Second Heap	Third Heap
8	16+12=28	12

It is clear now that before the first transposition (i.e., before we took as many matches from the first heap as there were in the second and added them to the second) the number of matches in each heap was:

First Heap	Second Heap	Third Heap
22	14	12

9. This riddle, too, can best be solved in reverse. We know that when the money was doubled for the *third* time, there was 1 ruble 20 kopeks in the pouch (that is the sum the old man received the last time). How much was there in the pouch before that? Obviously 60 kopeks. That was what remained after the peasant had paid the old man his second ruble and 20 kopeks. Therefore, before the payment there was:

$$1.20+0.60=1.80.$$

Further: 1 ruble 80 kopeks was the sum after the money had been doubled for the *second* time. Before that there were only 90 kopeks, i.e., what remained after the peasant had paid the old man his first ruble and 20 kopeks. Hence, before the first payment there were 0.90+1.20=2.10 in the pouch. That was after the first operation. Originally, therefore, there was half that amount, or 1 ruble 5 kopeks. That was the sum with which the peasant had started his unsuccessful get-rich-quick operation.

Let us verify:

	Money in the pouch:
After the first operation	1.05× 2 =2.10
After the first payment	2.10—1.20 =0.90
After the second operation	0.90× 2 =1.80
After the second payment	1.80—1.20 =0.60
After the third operation	0.60× 2 =1.20
After the third payment	1.20—1.20 =0

10. Our calendar comes from the early Romans who, before Julius Caesar, began the year in March. December was then the *tenth* month. When New Year was moved to January 1, the names of the months were not shifted. Hence the disparity between the meaning of the names of certain months and their sequence:

Month	Meaning	Place
September	(septem—seven)	9th
October	(octo—eight)	10th
November	(novem—nine)	11th
December	(deka—ten)	12th

11. Let us see what happened to the original number. First, a similar number was written alongside it. That is as if we took a number, multiplied it by 1,000 and then added the original number, e.g.:

$$872{,}872=872{,}000+872$$

It is clear that what we have really done was to multiply the original number by 1,001.

What did we do after that? We divided it successively by 7, 11 and 13, or by $7 \times 11 \times 13$, i.e., by 1,001.

So, we first multiplied the original by 1,001 and then divided it by 1,001. Simple, isn't it?

* * *

Before we close our chapter about the riddles at the holiday home, I should like to tell you of another three arithmetical tricks that you can try on your friends. In two you have to guess numbers and in the third, the owners of certain objects.

These tricks are very old and you probably know them well, but I am not sure at all people know what they are based on. And if you do not know the theoretic basis of tricks, you cannot expect to unravel them. The explanation of the first two will require an absolutely elementary knowledge of algebra.

12. THE MISSING DIGIT.—Tell your friend to write any multidigit number, but no ending in noughts, say, 847. Ask him to add up these three digits and then subtract the total from the original. The result will be:

$$847-19=828$$

Ask him to cross out any one of the three digits and tell you the remaining ones. Then you tell him the digit he has crossed out, although you know neither the original nor what your friend has done with it.

How is this explained?

Very simply: all you have to do is to find the digit which, added to the two you know, will form the nearest number divisible by 9. For instance, if in the number 828 he crosses out the first digit (8) and tells you the other two (2 and 8), you add them and get 10. The nearest number divisible by 9 is 18. The missing number is consequently 8.

How is that? No matter what the number is, if you subtract from it the total number of its digits, the balance will always be divisible by 9. Algebraically, we can take a for the number of hundreds, b for the number of tens and c for the number of units. The total number of units is therefore:

$$100a+10b+c.$$

From this number we subtract the sum total of its digits $a+b+c$ and we obtain:

$$100a+10b+c-(a+b+c)=99a+9b=9(11a+b)$$

But $9(11a+b)$ is, of course, divisible by 9. Therefore, when we subtract from a number the sum total of its digits, the balance is always divisible by 9.

It may happen that the sum of the digits you are told is divisible by 9 (for example, 4 and 5). That shows that the digit your friend has crossed out is either 0 or 9, and in that case you have to say that the missing digit is either 0 or 9.

Here is another version of the same trick: instead of subtracting from the original number the sum total of its digits, ask your friend to subtract the same number only transposed in any way he wishes. For instance, if he writes 8,247, he can subtract 2,748 (if the number transposed is greater than the original, subtract the original). The rest is done as described above: 8,247—2,748=5,499. If the crossed-out digit is 4, then knowing the other three (5, 9 and 9), you add them up and get 23. The nearest number divisible by 9 is 27. Therefore, the missing digit is 27—23=4.

13. WHO HAS IT?—This clever trick requires three little things that can be put in one's pocket—a pencil, a key and a penknife will do very well. In addition to that, place a plate with 24 nuts—draughts or domino pieces or matches will do just as well—on the table.

Having completed these preparations, ask each of your three friends to put one of the three things into his pocket—one the pencil, the second the key and the third the penknife. This they must do in your absence, and when you return to the room, you guess correctly where each object is.

The process of guessing is as follows: on your return (i.e, after each has concealed the object) you ask your friends to take care of some nuts—you give one nut to the first, two to the second and three to the third. Then you leave the room again, telling them that they must take more nuts—the one who has the pencil should take as many nuts as he was given the first time; the one with the key *twice* as many as he has been given; and the one with the penknife *four times* the number. The rest, you tell them, should remain in the plate.

When they have done that, they call you into the room. You walk in, look at the plate and announce what each of your friends has in his pocket.

The trick is all the more mystifying since you do it solo, so to speak, without any assistant who could signal to you secretly. There is really nothing tricky in the riddle—the whole thing is based on calculation. You guess who has each of the things from the number of nuts remaining in the plate. Usually there are not many of them left—from one to seven—and you can count them at one glance.

How, then, do you know who has what thing?

Simple. Each different distribution of the three objects leaves a different number of nuts in the plate. Here is how it is done.

Let us call your three friends Dan, Ed and Frank, or simply *D*, *E*, *F*. The three things will be as follows: the pencil—*a*, the key—*b*, and the penknife—*c*. The three objects can be distributed among the trio in just six ways:

D	*E*	*F*
a	*b*	*c*
a	*c*	*b*
b	*a*	*c*
b	*c*	*a*
c	*a*	*b*
c	*b*	*a*

There can be no other combinations—the table above exhausts all of them.

Now let us see how many nuts remain after each combination:

DEF	Number of Nuts Taken	Total	Remainder
abc	1+1=2; 2+4= 6; 3+12=15	23	1
acb	1+1=2; 2+8=10; 3+ 6= 9	21	3
bac	1+2=3; 2+2= 4; 3+12=15	22	2
bca	1+2=3; 2+8=10; 3+ 3= 6	19	5
cab	1+4=5; 2+2= 4; 3+ 6= 9	18	6
cba	1+4=5; 2+4= 6; 3+ 3= 6	17	7

You will see that the remainder is each time different. Knowing what it is, you will easily establish who has what in his pocket. Once again, for the third time, you leave the room, look at your notebook into which you have written the table above (frankly speaking, you need only the first and last columns). It is difficult to memorize this table, but then there is really no need for that. The table will tell where each thing is. If, for example, there are five nuts remaining in the plate, the combination is *bca*, i.e.,

Dan has the key,
Ed has the penknife, and
Frank has the pencil.

If you want to succeed, you must remember how many nuts you gave each of your three friends (the best way is to do so in alphabetic order, as we have done here).

CHAPTER II

MATHEMATICS IN GAMES

DOMINOES

14. A CHAIN OF 28 PIECES.—Can you lay out the 28 domino pieces in a chain, observing all the rules of the game?

15. THE TWO ENDS OF A CHAIN.—The chain of the 28 dominoes begins with five dots. How many dots are there on the other end of the chain?

16. A DOMINO TRICK.—Your friend takes a domino piece and asks you to build a chain with the remaining 27, affirming that this can be done no matter what piece is missing. After that he leaves the room.

You lay out the dominoes in a chain, and find that your friend is right. What is even more wonderful is that your friend, without seeing your chain, can tell you the number of dots on each of the end pieces.

How can he know that? And why is he so certain that a chain can be built up of any 27 pieces?

17. A FRAME.—Fig. 5 shows a square frame made up of domino pieces in accordance with the rules of the game. The sides are equal in length, but not in the total number of dots. The top and left sides add up to 44 points each, while the other two to 59 and 32 points, respectively.

Can you build a square frame in which each side will have 44 points?

18. SEVEN SQUARES.—It is possible to build a four-domino square in such a way as to have the same number of dots on each side, as shown in fig 6: there are 11 dots on each side.

Can you make *seven* such squares out of the 28 domino pieces? It is not necessary for all the sides of the seven squares to have an identical number of dots, — just the four sides of each square.

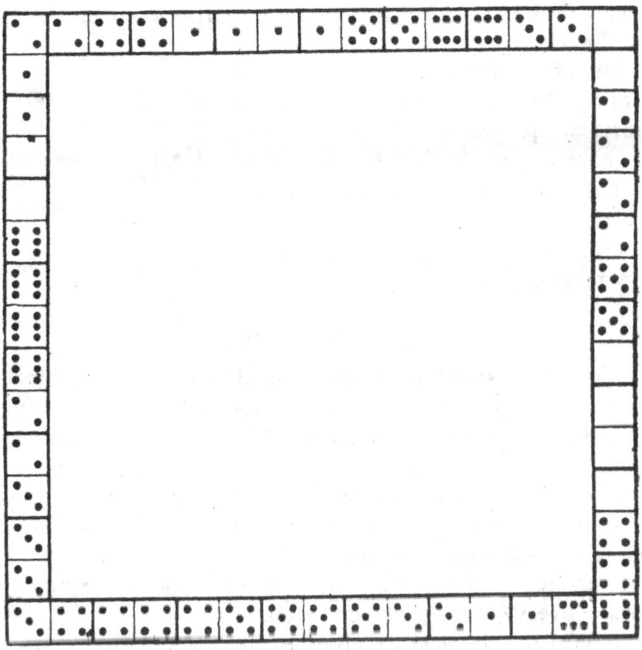

Fig. 5. A domino frame.

19. MAGIC SQUARES.—Fig. 7 depicts a square of 18 domino pieces. The wonder of it is that there are 13 dots in every one of its rows—vertical, horizontal or diagonal. From time immemorial these squares have been called "magic."

Arrange several other 18-piece magic squares, but with a different number of dots. Thirteen is the lowest total in a row and 23 the highest.

20. PROGRESSION IN DOMINOES.—Fig. 8 shows six dominoes arranged according to the rules of the game, with the number of dots increasing by one on each successive piece: four on the first, five on

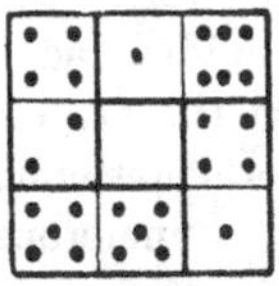

Fig. 6. A domino square.

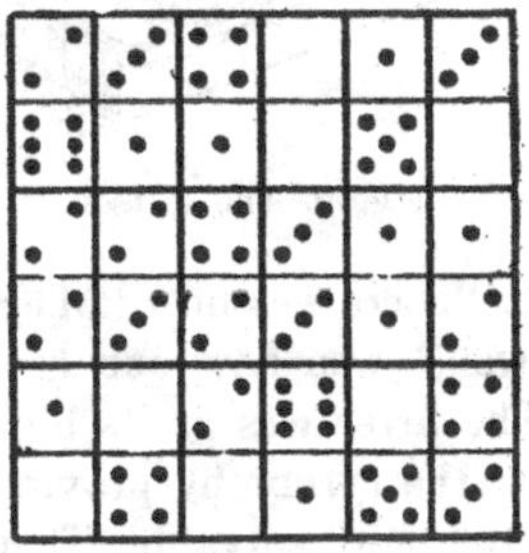

Fig. 7. A magic square.

the second, six on the third, seven on the fourth, eight on the fifth and nine on the sixth.

A series of numbers increasing (or decreasing) by the same amount is known as "arithmetical progression." In our case, each number is greater than the preceding by one, but there can be any other "difference."

The task is to arrange several other six-piece progressions.

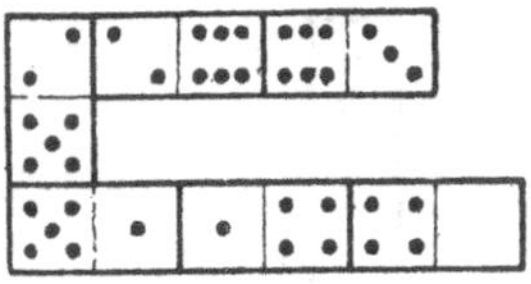

Fig. 8. The Fifteen Puzzle.

THE FIFTEEN PUZZLE

The story of the well-known square shallow box with 15 blocks numbered 1 to 15 inclusive is an extremely interesting one, though very few players know it. Here is what W. Ahrens, German mathematician and draughts expert, wrote about it:

"In the late 1870's there appeared in the United States a new game —'The Fifteen Puzzle.' Its popularity spread fast and wide, and it soon became a real social calamity.

"The craze hit Europe too. One came across people trying to solve the puzzle everywhere—even in public conveyances. Office workers and shop salesmen became so absorbed in working it out that their employers, driven to desperation, had to forbid the game during working hours. Enterprising people took advantage of the mania to arrange large-scale tournaments. The puzzle made its way even into the German Reichstag. The well-known geographer and mathematician Siegmund Günther, a Reichstag deputy at the time of the craze, recalled seeing his grey-haired colleagues bending thoughtfully over the little square boxes.

Fig. 9. The Fifteen Puzzle.

"In Paris the game was played in the open air, on the boulevards and soon spread from the capital to the provinces. 'There was no rural homestead where this spider had not woven its web,' was how one French author described the craze.

"The fever was at its highest in 1880, but mathematicians soon defeated the tyrant by proving that only half of the numerous problems it posed were solvable. There was absolutely no chance of finding a solution for the rest.

"The mathematicians made it clear why some problems remained unsolved despite all efforts and why the organizers of tournaments were not afraid of offering huge prizes for their solution. In this the inventor of the puzzle, Sam Lloyd, surpassed everyone. He asked a New York newspaper owner to offer $1,000 to anyone who would solve a certain variant of the Fifteen Puzzle, and when the publisher hesitated, Lloyd said he would pay the sum himself. Lloyd was well known for his clever conundrums and brain-teasers. Curiously enough, he could not get a U.S. patent for his puzzle. According to regulations, a person applying for one was required to submit a 'working model.' At the Patent Office he was asked if the puzzle was solvable, and Lloyd had to admit that mathematically it was not. 'In that case,' the official said, 'there can be no working model and without it there can be no patent.' Lloyd left it at that, but there is no doubt that he would have been much more insistent could he have but foreseen the unusual success of his invention."

Here are some facts about the puzzle, as told by the inventor himself:

"Puzzle enthusiasts may well remember how, in the 1870's, I caused the world to rack its brain over a box with moving blocks, which became known as 'The Fifteen Puzzle.' Thirteen of the blocks were ar-

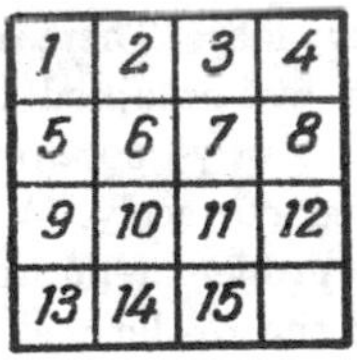

Fig. 10. Normal order of blocks (position I).

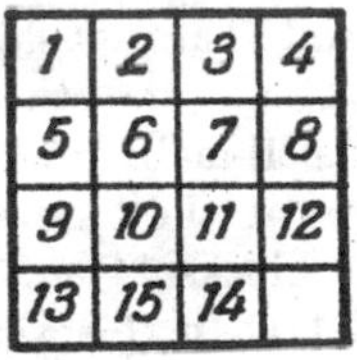

Fig. 11. Insoluble (position II).

ranged in regular order and only two, 14 and 15, were not (see fig. 11). The task was to shift one block at a time until blocks 14 and 15 were brought into regular order.

"No one won the $1,000 prize offered for the first correct solution, although people worked tirelessly at it. There are many humorous stories told of tradesmen who were so absorbed in the puzzle they forgot to open their shops and of respected officials who spent nights seeking for a way to solve the problem. People just would not give up their search for the solution, being confident of success. Navigators ran their ships against reefs, locomotive engineers missed stations and farmers chucked up their ploughs."

* * *

We shall acquaint the reader with the rudiments of the puzzle. On the whole, it is extremely complicated and closely connected with one of the sections of advanced algebra ("the theory of determinants"). Here is what Ahrens wrote about it:

"The job is to shift the blocks by using the blank space in such a manner as to finally arrange all the 15 in regular order, i.e., to have block 1 in the upper left-hand corner, block 2 to the right of it, block 3 next to block 2, and block 4 in the upper right-hand corner; to have blocks 5, 6, 7 and 8 in this order in the next row, etc. (see fig. 10).

"Imagine for a moment that the blocks are all placed at random. It is always possible to bring block 1 to its correct position through a series of moves.

"It is equally possible to shift block 2 to the next square without touching block 1. Then, without touching block 1 and 2, one can move blocks 3 and 4 into their places. If, perchance, they are not in the last two vertical rows, it is easy to bring them there and to achieve the desired result. The top row 1, 2, 3 and 4 is now in order and in our subsequent operations we shall leave these four blocks alone. In the same way we shall try to put in order the row 5, 6, 7 and 8 ; this is also possible. Further, in the next two rows it is necessary to bring blocks 9 and 13 to their correct positions. Once in order, blocks 1, 2, 3, 4, 5, 6, 7, 8, 9 and 13 are not shifted any more. There remain six squares —one of them blank and the other five occupied by blocks 10, 11, 12, 14 and 15 in pell-mell order. It is always possible to shift blocks 10, 11 and 12 until they are arranged correctly. When this is done, there will remain blocks 14 and 15 in proper or improper order in the bottom row (fig. 11). In this way—as the reader can verify himself—we come to the following result:

"Any original combination of blocks can be brought into the order shown in fig. 10 (position I) or fig. 11 (position II).

"If a combination, which we shall call *C* for short, can be rearranged into position I, then it is obvious that we can do the reverse too, i. e., rearrange position I into combination *C*. After all, every move can be reversed: if, for instance, we can shift block 12 into the blank space, we can bring it back to its old position just as well.

"Thus, we have two series of combinations: in the first we can bring the blocks into regular order (position I) and in the second into position II. And conversely, from regular order we can obtain any combination of the first series and from position II any combination of the second series. Finally, any one of the two combinations of the same series may be reversed.

"Is it possible to transform position I into position II? It may definitely be proved (without going into detail) that no number of moves is capable of doing that. Therefore, the huge number of combinations of blocks may be classed into two series—the first series where the blocks can be arranged in regular order, i.e., the solvable; and the second series where in no circumstances can the blocks be brought into regular order, i.e., the insoluble, and it is for the solution of these positions that huge prizes were offered.

"Is there any way of telling to what series the position belongs. There is, and here is an example.

"Let us analyze the position in fig. 12. The first row of blocks is in order and the second, too, with the exception of block 9. This block occupies the space that rightfully belongs to block 8. Therefore, block 9 precedes block 8. Such violation of regular order is termed disorder.' Our analysis further shows that block 14 is three spaces ahead of its regular position, i.e., it precedes blocks 11, 12 and 13. Here we have three 'disorders' (14 before 12, 14 before 13, and 14 before 11). Altogether we have 1+3=4 'disorders.' Further, block 12 precedes block 11, just as block 13 precedes block 11. That gives us another two 'disorders' and brings their total to 6. In this way we determine the number of 'disorders' in each position, taking good care to vacate beforehand the lower right-hand corner. If the total number of 'disorders,' as in this case, is *even*, then the blocks can be arranged in regular order and the problem is solvable. If, on the other hand, the number of 'disorders' is *odd*, then the position belongs to the second category, i.e., it is insoluble.

"Mathematical explanation of this puzzle has dealt a death blow to the craze. Mathematics has created an exhaustive theory of the game, leaving no place whatever for doubts. The solution of the puzzle does not depend on guesswork or quick wit, as in other games, but solely on mathematical factors that predetermine the result with absolute certainty."

Let us now turn to some problems in this field.

Below are three of the *solvable* ones, made up by the inventor.

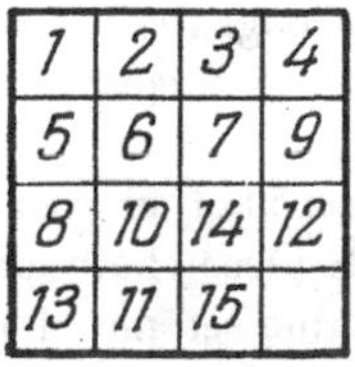

Fig. 12. The blocks are not in order.

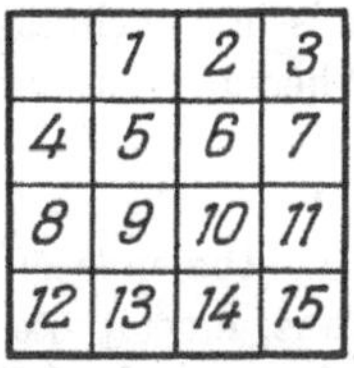

Fig. 13. Lloyd's first problem.

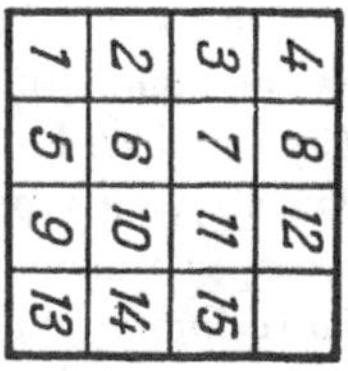
Fig. 14. Lloyd's second problem.

21. THE FIRST PROBLEM.—In fig. 12 arrange the blocks in regular order, leaving the upper left-hand corner blank (as in fig. 13).

22. THE SECOND PROBLEM.—Take the box as in fig 10, place it on its side (one-quarter turn) and shift the blocks until they assume the positions in fig. 14.

23. THE THIRD PROBLEM.—Shifting the blocks according to the rules of the puzzle, turn the box into a "magic square," i.e., arrange the blocks in such a manner as to obtain a total of 30 in all directions.

Answers 14 to 23

14. To simplify the problem let us set aside all the seven double pieces: 0—0, 1—1, 2—2, etc. There will remain 21 pieces with each number repeated six times. For instance, there will be four dots (on one half of the piece) on the following six pieces:

4—0, 4—1, 4—2, 4—3, 4—5, and 4—6.

We thus see that each number is repeated an *even* number of times. It is obvious that these pieces can be linked up. And when this is done, when the 21 pieces are arranged in an uninterrupted chain, then we insert the seven double pieces between pieces ending with the same number of dots, i.e, between two 0's, two 1's, two 2's, etc. After that all the 28 pieces will be drawn into the chain according to the rules of the game.

15. It is easy to prove that a chain of 28 pieces must end with the same number of dots as it begins. Indeed, if it were not so, the number of dots on the ends of the chain would be repeated an *odd* number of times (inside the chain the numbers always lie in pairs). However, we know that in a complete set each number is repeated eight times, i.e., an even number of times. Therefore, our assumption about the unequal number of dots on the ends of the chain is wrong: the number of dots must be the same (in mathematics arguments of this sort are called "reductio ad absurdum").

Incidentally, this property of the chain has another extremely interesting aspect: namely, that the ends of a 28-piece chain can always be linked to form a ring. Thus, a complete domino set may be arranged according to the rules of the game both into a chain with loose ends or into a ring.

The reader may be interested to know how many ways there are of arranging this ring or chain. Without going into tiresome calculations we can say that there is a gigantic number of ways of doing this. It is exactly 7,959,229,931,520 (representing the value of $2^{13}\times 3^{8}\times 5\times 7\times 4{,}231$).

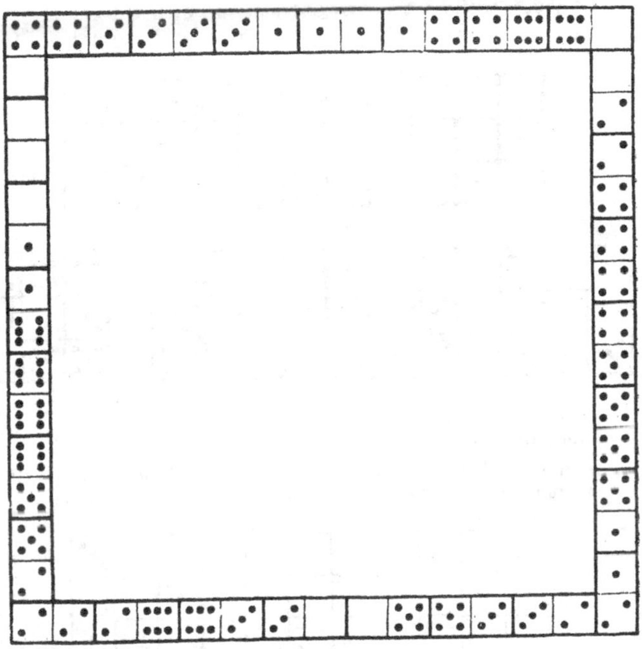

Fig. 15.

16. The solution of this problem is similar to the one described above. We know that the 28 domino pieces can always be arranged to form a ring. Therefore, if we take one piece away,

1) the remaining 27 will always form an uninterrupted chain with loose ends;
2) the numbers on the loose ends of this chain will always be those on the two halves of the piece taken away.

Concealing a domino piece, you can always tell beforehand the number of dots on each end of the chain.

17. The total number of dots on the four sides of the unknown square must equal $44\times 4=176$, i.e., 8 more than the total number of dots on all the domino pieces (168). This is because the numbers at the vertices of the square are counted twice. This determines that the total of the numbers at the vertices is 8 and that helps to find the necessary

arrangement (although its discovery nevertheless remains quite troublesome. The solution is shown in fig. 15).

18. Here are two of the many solutions of this problem. In the first (fig. 16) we have:

1 square with a total of 3	2 squares with a total of 9
1 square with a total of 6	1 square with a total of 10
1 square with a total of 8	1 square with a total of 16

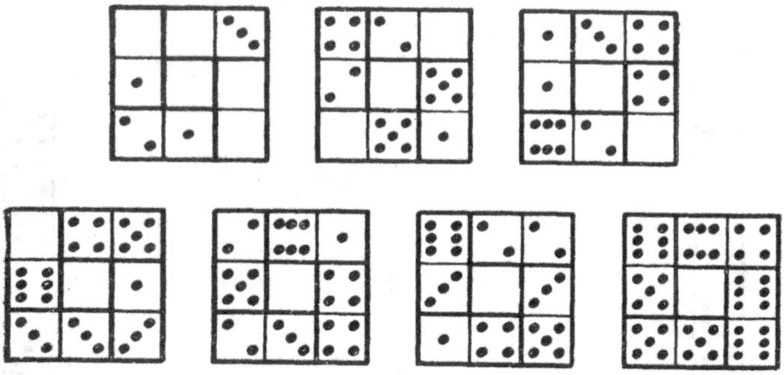

Fig. 16.

In the second (fig. 17) we have:

2 squares with a total of 4	2 squares with a total of 10
1 square with a total of 8	2 squares with a total of 12.

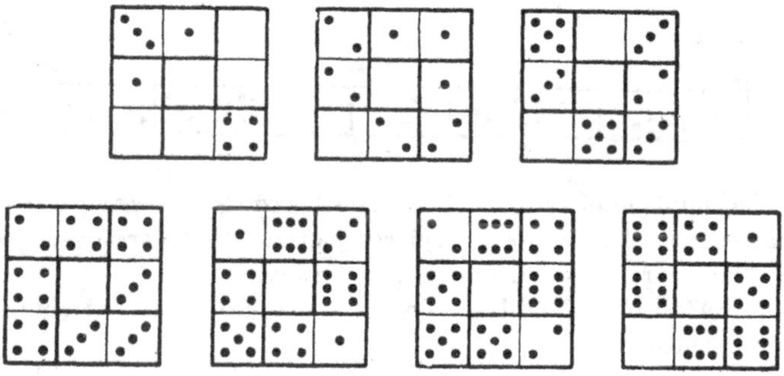

Fig. 17.

19. Fig. 18 is a specimen of a magic square with a total of 18 dots in each row.

20. Here, as an example, are two progressions with a difference of 2:

a) 0—0, 0—2, 0—4, 0—6, 4—4 (or 3—5), 5—5 (or 4—6).

b) 0—1, 0—3 (or 1—2), 0—5 (or 2—3), 1—6 (or 3—4), 3—6 (or 4—5), 5—6.

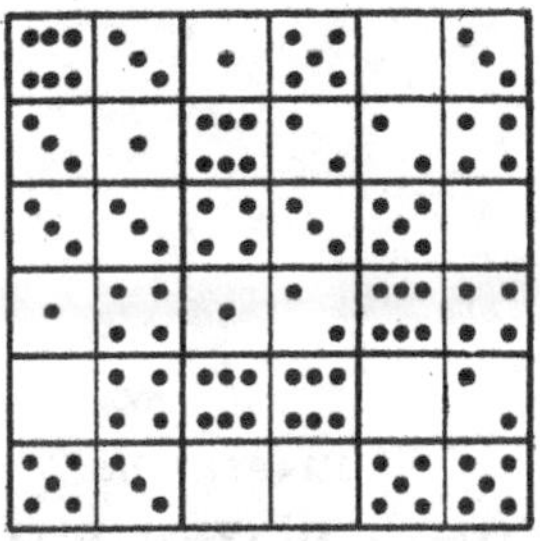

Fig. 18.

There are altogether 23 six-piece progressions. The starting pieces are:

a) for progressions with a difference of 1:

0—0	1—1	2—1	2—2	3—2
0—1	2—0	3—0	3—1	2—4
1—0	0—3	0—4	1—4	3—5
0—2	1—2	1—3	2—3	3—4

b) for progressions with a difference of 2:

0—0 0—2 0—1

21. This problem may be solved by the following 44 moves:

14, 11, 12, 8, 7, 6, 10, 12, 8, 7
4, 3, 6, 4, 7, 14, 11, 15, 13, 9
12, 8, 4, 10, 8, 4, 14, 11, 15, 13,
9, 12, 4, 8, 5, 4, 8, 9, 13, 14,
10, 6, 2, 1

22. This problem may be solved by the following 39 moves:

14, 15, 10, 6, 7, 11, 15, 10, 13, 9,
5, 1, 2, 3, 4, 8, 12, 15, 10, 13,
9, 5, 1, 2, 3, 4, 8, 12, 15, 14,
13, 9, 5, 1, 2, 3, 4, 8, 12

23. The magic square with the total of 30 is achieved through the following moves:

12, 8, 4, 3, 2, 6, 10, 9, 13, 15,
14, 12, 8, 4, 7, 10, 9, 14, 12, 8,
4, 7, 10, 9, 6, 2, 3, 10, 9, 6,
5, 1, 2, 3, 6, 5, 3, 2, 1, 13,
14, 3, 2, 1, 13, 14, 3, 12, 15, 3

CHAPTER III

ANOTHER DOZEN PUZZLERS

24. STRING.—"What! More string?" the boy's mother exclaimed, tearing herself away from washing. "D'you think I'm made of it? All I hear is, 'Give me some string.' I gave you a whole ball yesterday. What d'you need so much for? What have you done with the one I gave you?"

"What have I done with it?" the boy countered. "First, you took half of it back."

"And how d'you expect me to tie the washing?"

"Then Tom took half of what remained to fish stickle-backs in the creek."

"Yes, you couldn't very well refuse your elder brother."

"I didn't. There remained very little and Dad took half of it to fix his suspenders. And then sis took two-fifths to tie her braids. . . . "

"And what have you done with the rest?"

"With the rest? There were only 30 centimetres left. Try to play telephone with that!"

How much string was there in the first place?

25. SOCKS AND GLOVES.—In one box there are 10 pairs of brown and 10 pairs of black socks and in another the same amount of brown and black gloves. How many socks and gloves must one take out of the boxes to select one pair of socks and one pair of gloves—of the same colour, of course?

26. LONGEVITY OF HAIR.—How many hairs are there on the average on a man's head? About 150,000.* It has been calculated that a man sheds about 3,000 hairs a month.

Given this, can you calculate the longevity—average, of course—of each hair on a man's head?

---

* Many may wonder how we come by this figure. Did we have to count the hairs? No. It is enough to count them on one square centimetre of man's head. Knowing this and the size of the hair-covered surface, it is not hard to determine the total. Anatomists use the method resorted to by sylviculturists in counting trees in a forest.

27. WAGES.—Together with overtime my wages last week were 250 rubles. My basic wages are 200 rubles above overtime. How much do I earn without overtime?

28. SKIING.—A man has calculated that if he skis 10 kilometres an hour, he will arrive at a certain place at 1 p.m.; if he does 15 kilometres, he will reach the same spot at 11 a.m. How fast must he ski to get there at 12 noon?

29. TWO WORKERS.—Two workers, one old and the other young, live in the same house and work at the same factory. It takes the young man 20 minutes to walk to the plant. The old man covers the distance in 30 minutes. When will the young worker catch up with the older man if the latter starts out five minutes before him?

30. TYPING.—Two girls are asked to type a report. The more experienced can do the whole job in two hours; the other in three.

How long will it take them to finish the whole job if they divide it in such a manner as to complete it as early as possible?

Problems of this type are usually solved as follows: we find what part of the job each does in an hour, add the two parts and divide 1 by the result. Can you think up a new way of solving this problem?

31. TWO COG-WHEELS.—An eight-tooth cog-wheel is coupled with a 24-tooth one (fig. 19). How many times must the small one rotate on its axis to circle around the big one?

32. HOW OLD IS HE?—A conundrum enthusiast was asked how old he was. The reply was quite ingenious.

"Take my age three years hence, multiply it by 3 and then subtract three times my age three years ago and you will know how old I am."

Well, how old is he?

33. ANOTHER AGE RIDDLE.—"How old is Ivanov?" a friend of mine asked me the other day.

"Ivanov? Let's see. Eighteen years ago he was *three times* as old as his son."

"But he's only *twice* as old now," my friend interrupted.

"That's right and this is why it isn't difficult to arrive at their ages."

Well, reader?

Fig. 19. How many times must the small cog-wheel rotate?

34. SHOPPING.—I had about 15 rubles in one-ruble notes and 20-kopek coins when I went out shopping. When I returned, I had as many one-ruble notes as I originally had

20-kopek coins and as many 20-kopek coins as I originally had one-ruble notes. Briefly, I came back with about one-third of what I had started out with.

How much did I spend?

Answers 24 to 34

24. When the boy's mother took half of the string, there naturally remained $^1/_2$. After his brother there remained $^1/_4$, after his father, $^1/_8$ and after his sister, $^1/_8 \times ^3/_5 = ^3/_{40}$. If 30 cm. $= ^3/_{40}$, then the original length was $30 : ^3/_{40} = 400$ cm. or 4 m.

25. It is enough to take three socks, for two of them will always be of the same colour. It is not so simple with gloves, for they differ not only in colour, but also in that half of them are for the right hand and the rest for the left. Here you must take at least 21 gloves. If you take less, say, 20, they may all be for the left hand (ten brown and ten black).

26. The hair that falls last is the one that is the youngest today, i.e., the one that is only one day old.

Let us compute how long it will take before the last hair falls. In the first month a man sheds 3,000 hairs out of the 150,000 he has on his head; in the first two months 6,000; and in the first year $3{,}000 \times 12 = 36{,}000$. Therefore, it will take a little over four years for the last hair to fall. It is thus that we have determined the average age of human hair.

27. Many say 200, without even stopping to think. That is wrong, for then the basic wages would be only 150 rubles and not 200.

Here is how the problem should be solved. We know that if we add 200 rubles to overtime we get the basic wages. Therefore, if we add 200 rubles to 250 rubles we have two basic wages. But $250+200=450$. That means two basic wages equal 450 rubles. Hence, my wages without overtime amount to 225 rubles and overtime to 25 rubles.

Let us verify: $225-25=200$. And that is as the problem has it.

28. This problem is interesting for two reasons. First, it may easily lead one to think that the speed we seek is the mean result of 10 and 15 kilometres, i.e., 12.5 kilometres an hour. It is not hard to guess that this is wrong. Indeed, if the distance the skier covers is *a* kilometres, then going at 15 kilometres an hour he will require $\frac{a}{15}$ hours to cover it, while going at 10 kilometres $\frac{a}{10}$, and at 12.5 kilometres $\frac{a}{12^1/_2}$ or $\frac{2a}{25}$. Thus, the equation:

$$\frac{2a}{25}-\frac{a}{15}=\frac{a}{10}-\frac{2a}{25}$$

because each of these members is equal to one hour.

Simplifying by a, we obtain

$$\frac{2}{25}-\frac{1}{15}=\frac{1}{10}-\frac{2}{25}$$

or the arithmetical proportion

$$\frac{4}{25}=\frac{1}{15}+\frac{1}{10}$$

This equation is wrong because

$$\frac{1}{15}+\frac{1}{10}=\frac{1}{6}, \text{ i.e., } \frac{4}{24} \text{ and not } \frac{4}{25}$$

The other reason why it is interesting is because it can be solved orally, without equations.

Here is how it goes: if the skier did 15 kilometres an hour and was out for two hours more (i.e., as long as if he were skiing at 10 kilometres an hour), he would cover an additional 30 kilometres. In one hour, we know, he covers 5 km. more. Thus, he would be out for 30 : 5= 6 hours. This determines the duration of the run at 15 kilometres an hour: 6—2=4 hours. And it is not hard now to find the distance covered: 15×4=60 kilometres.

Now it is easy to see how fast he must ski to arrive at that place at 12 noon, i. e., in five hours:

$$60 : 5 = 12 \text{ kilometres.}$$

It is not difficult to verify the correctness of the answer.

29. This problem may be solved in many ways without equations.

Here is the first way. In five minutes the young worker covers $\frac{1}{4}$ of the way and the old $\frac{1}{6}$, i.e., $\frac{1}{4}-\frac{1}{6}=\frac{1}{12}$ less than the young man.

Since the old man was $\frac{1}{6}$ of the way ahead of the young worker, the latter would catch up with him after $\frac{1}{6}:\frac{4}{12}=2$ five-minute intervals, or 10 minutes.

The other way is even simpler. To get to the factory the old worker needs 10 minutes more than the young one. If he were to leave home 10 minutes earlier, they would both arrive at the plant at the same time. If the old worker were to leave only 5 minutes earlier, the young man would overhaul him half-way to the factory, i.e., 10 minutes later (since it takes him 20 minutes to cover the whole distance).

There are other arithmetical solutions too.

30. A novel way of solving this problem is as follows: let us find out how the typists should divide the job to finish it at the same time (it is evident that this is the only way to finish work as quickly as possible, provided, of course, they do not idle their time away). Since the more experienced typist can work $1\frac{1}{2}$ times faster that the other, t is clear that her share should be $1\frac{1}{2}$ times greater. Then both will finish simultaneously. Hence, the first should take $^3/_5$ of the report and the second $^2/_5$.

Generally speaking, this solves the problem. There now remains only to find how long it takes the first typist to do her share, i.e., $^3/_5$ of the report. We know that she can do the whole job in 2 hours. Therefore, $^3/_5$ will be done in $2\times{}^3/_5={1}^1/_5$ hours. The other typist must finish her bit in the same span.

Thus, the fastest time the two can finish the job is 1 hour 12 minutes.

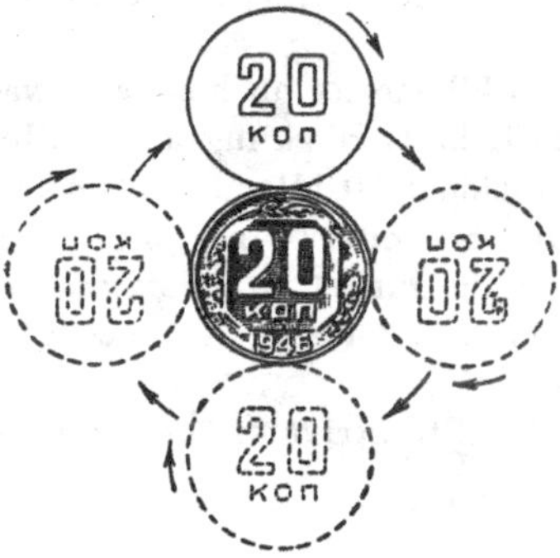

Fig. 20.

31. If you think the small cogwheel will rotate three times, you are very much mistaken. It is four times.

To see why this is so, take a sheet of paper and place on it two equal-sized coins—two 20-kopek coins will do (fig. 20). Then, holding the lower one tight in its place, roll the upper coin around it. You will be surprised to see that by the time the upper coin reaches the bottom of the lower one it will have fully rotated on its axis. This may be seen from the position of the denomination figures stamped on the coin. And when it has done a complete circle around the lower coin, it will have rotated twice.

Generally speaking, when a body rotates round a circle, it always does one revolution more than one can count. It is precisely this that explains why the earth, revolving round the sun, succeeds in rotating on its axis not in $365\frac{1}{4}$ days, but in $366\frac{1}{4}$ days, if one counts its revolutions in respect to the stars and not the sun. You will understand now why sidereal days are shorter than solar days.

32. Arithmetically, the solution of this problem is quite complicated, but it becomes simple when we apply algebra. Let us take x for the years. The age three years hence will be $x+3$, and the age three years ago $x-3$. We thus have the equation:

$$3(x+3)-3(x-3)=x$$

Solving this we obtain $x=18$. The conundrum enthusiast is 18 years old.

Let us verify: Three years hence he will be 21; three years ago he was 15.

The difference is

$$(3\times 21)-(3\times 15)=63-45=18$$

33. Like the preceding problem, this one is also solved by simple equation. If the son is x years old, then the father is $2x$ years old. Eighteen years ago they were both 18 years younger: the father was $2x-18$ and the son $x-18$. It is known that the father was then three times as old as the son:

$$3(x-18)=2x-18$$

Solving this equation, we find that x equals 36. The son is 36 and the father 72.

34. Let us assume that originally I had x rubles and y 20-kopek coins.

Going shopping, I had

$$(100x+20y) \text{ kopeks}$$

I returned with only

$$(100y+20x) \text{ kopeks}$$

This last sum, as we know, is one-third of the original. Therefore

$$3(100y+20x)=100x+20y$$

Simplifying, we have

$$x=7y$$

If $y=1$, then $x=7$. Assuming this is so, I had 7.20 rubles when I set out shopping. This is wrong, for the problem says I had "about 15 rubles."

Let us see what we get if $y=2$. Then $x=14$. The original sum was 14.40 rubles, which accords with the conditions of the problem.

If we assume that $y=3$, then the sum will be too big—21.60 rubles.

Therefore, the only suitable answer is 14.40 rubles. After shopping I had two one-ruble notes and 14 twenty-kopek coins, i.e., $200+280=480$ kopeks. That is indeed one-third of the original sum ($1{,}440 : 3=480$).

My purchases, therefore, cost $14.40-4.80=9.60$ rubles.

CHAPTER IV

COUNTING

35. DO YOU KNOW HOW TO COUNT?—Anyone over three years of age will probably consider himself insulted if he is asked that. Indeed, one requires no skill whatever to say 1, 2, 3, etc. And yet I am sure that sometimes you find counting rather complicated. It all depends, of course, on what you have to count. For instance, it is not difficult to count nails in a box. But just suppose that apart from the nails the box also contains a number of screws and you are asked to find out how many of each there are. What will you do then? Separate the nails from the screws and then count them?

That is the task women often face when they take washing to a laundry. They have to itemize everything and to do that they have to sort out shirts, towels, pillow-cases, etc. Having done this rather tiresome job, they proceed to count them.

If that is how you count things, then you do not know how to count. This method is inconvenient, bothersome and at times plainly impracticable. It isn't half bad if you have to count nails or clothes, they can easily be sorted out. But just imagine you are a forester and are asked how many pines, firs, birches and aspens there are on each hectare. Well, here it is impossible to sort them out or group them by family. What are you going to do: count the pines, birches, firs and aspens separately? If you do that, you will have to walk around the forest four times.

There is an easier way of doing it—*in just one go*. I shall show you how this is done with nails and screws.

To count nails and screws in a box without sorting them out, you need, first of all, a pencil and a sheet of paper lined out as follows:

Nails	Screws

Fig. 21. Strokes must be jotted down in fives.

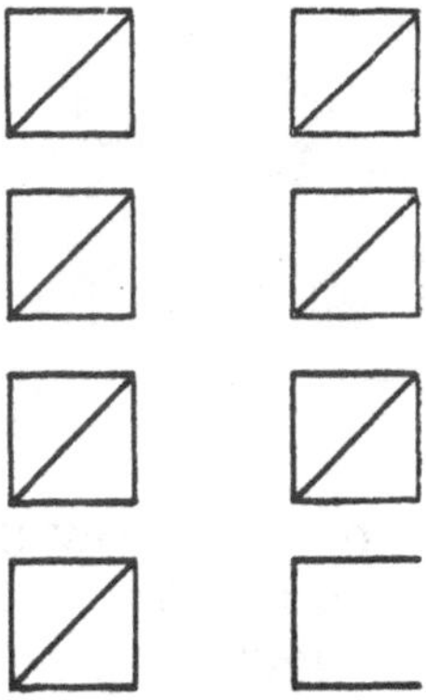

Fig. 22. How to count with squares.

Then you start counting. You take a thing out of the box and if it is a nail you put a stroke in the appropriate column. You do the same in the case of a screw, and thus continue until there is nothing left in the box. In the end you will have as many strokes in the "nail" column as there were nails in the box and ditto for the "screw" column. After that, all you have to do is add them up.

The addition of these strokes may be simplified and speeded up if you jot them down in fives in the form of little squares (fig. 21).

Squares of this sort are best grouped in pairs, i.e., after the first ten strokes jot down the eleventh in a new column. When there are two squares in the second column, start the third, etc. You will then have your strokes as shown in fig. 22.

It is very easy to count them, for you will see right away that there are three lots of 10 strokes each, one square of 5 and one incomplete figure of 3 strokes, i.e., 30+5+3=38.

You can use other figures too. For instance, a full square is often used to represent 10 (fig. 23).

In counting trees of different families you follow the same rule, only in this case you will have, say, four columns instead of two. It is also more convenient to have horizontal and not vertical columns. Take fig. 24 below as a specimen.

Fig. 25 shows what this form will look like when filled in.

After that it is very easy to find the total of each column:

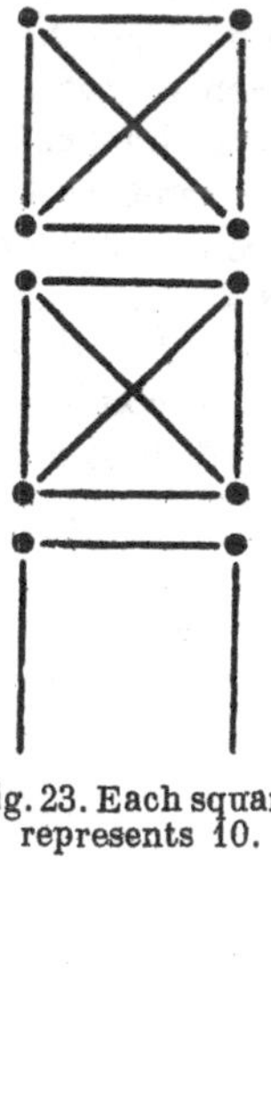

Fig. 23. Each square represents 10.

Pines	53
Firs	79
Birches	46
Aspens	37

This is also the method employed by medical workers in counting red and white corpuscles in a drop of blood.

Pines	
Firs	
Birches	
Aspens	

Fig. 24. The form for counting trees.

Women can save a lot of time and labour by adopting this method in itemizing their washing.

Now you know how best to count different plants growing on a plot. You draw a form, writing down each different plant in a different column, leaving a few columns in reserve for any other plants you may come across, and then start counting. A specimen form is given in fig. 26.

Then you proceed in exactly the same way as when you counted trees in the forest.

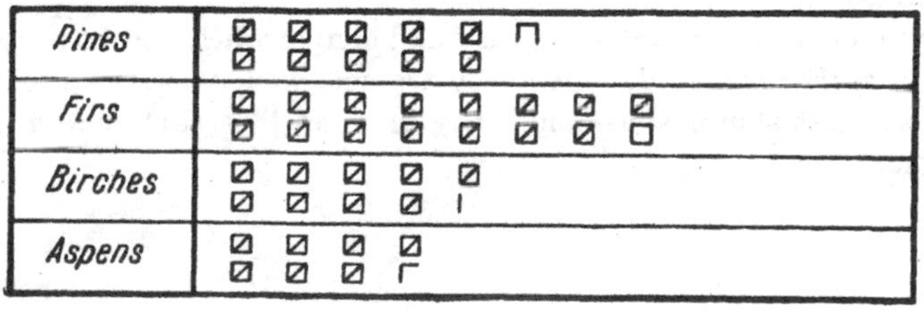

Fig. 25. How the form looks when filled in.

36. WHY COUNT TREES IN A FOREST?—Indeed, why? City dwellers, as a rule, think this is impracticable. In Lev Tolstoi's *Anna Karenina*, Levin, who is quite a farmer, talks with Oblonsky, who is about to sell a forest.

"Have you counted the trees?" he asks the latter.

Dandelions	
Buttercups	
Plantains	
Easter Bells	
Shepherd's-purses	

Fig. 26. How to count plants.

"What? Count my trees?" Oblonsky is surprised. "Count the sand on the seashore, count the rays of the planets—though a lofty genius might. . . ."

"Well," Levin interrupts him. "I tell you the lofty genius of Ryabinin succeeded. A merchant never purchases without counting."

People count trees in a forest to determine how many cubic metres of timber there are. To do that, they do not count all the trees, just part of them, say on 0.25 or 0.5 hectare, taking care to choose a place with average density of growth and average size of trees. For this, one must, of course, have an experienced eye. It is not enough simply to know how many trees of each family there are. It is also necessary to know how thick they are. Thus, the form will probably have more than the four columns we give in our simplified version. You may well imagine how many times we would have to walk around the forest to count the trees in the ordinary way and not in the way we have explained here.

As you may see, counting is easy and simple when you have to count things *of the same kind*. When they are not, we have to use the method we have just shown you—and many have no idea that such a method exists.

CHAPTER V

BAFFLERS WITH NUMBERS

37. HUNDRED RUBLES FOR FIVE.—A stage magician once made the following attractive proposal to his audience.

"I shall pay 100 rubles to anyone who gives me 5 rubles in 20 coins—50-kopek, 20-kopek and 5-kopek coins. One hundred for five! Any takers?"

The auditorium was silent. Some people armed themselves with paper and pencil and were evidently calculating their chances. No one, it seemed, was willing to take the magician at his word.

"I see you find it too much to pay 5 rubles for 100," the magician went on. "All right. I'm ready to take 3 rubles in 20 coins and pay you 100 rubles for them. Queue up!"

But no one wanted to queue up. The spectators were slow in taking up this chance of making "easy" money.

"What?! You find even 3 rubles too much. Well, I'll reduce it by another ruble—2 rubles in 20 coins. How's that?"

And still there were no takers. The magician continued:

"Perhaps you haven't any small change? It's all right. I'll trust you. Just write down how many coins of each denomination you'll give me."

On my part, I promise to pay 100 rubles to each reader who sends me such a list of coins.

38. ONE THOUSAND.—Can you write 1,000 by using eight identical digits (in addition to digits you may use signs of operation)?

39. TWENTY-FOUR.—It is very easy to write 24 by using three 8's: $8+8+8$. Can you do that by using three other identical digits? There is more than one solution to this problem.

40. THIRTY.—The number 30 may easily be written by three 5's: $5\times5+5$. It is harder to do it by using three other identical digits. Try it. You may find several solutions.

41. RESTORATION OF DIGITS.—In the following multiplication more than half of the digits are expressed by x's.

```
   ×1×
   3×2
   ---
   ×3×
 3×2×
×2×5
------
 1×8×30
```

Can you restore the missing digits?

42. WHAT ARE THE DIGITS?—Here is another similar problem. The task is to find the missing digits.

```
   ××5
   1××
  -----
  2××5
  13×0
 ×××
 ------
  4×77×
```

43. DIVISION.—Restore the missing digits in the following problem:

```
 ×2×5×|325
×××    1××
-----
 ×0××
 ×9××
-----
  ×5×
  ×5×
  ---
```

44. DIVIDING BY 11.—Write a number of nine non-repetitive digits that is divisible by 11.

Write the biggest of such numbers and then the smallest.

45. TRICKY MULTIPLICATION.—Look carefully at the following example:

$$48 \times 159 = 7{,}632$$

The interesting thing is that all the nine digits are different.

Can you give several other similar examples? If they do exist at all, how many of them are there?

46. A NUMBER TRIANGLE.—Write the nine non-repetitive digits in the circles of this triangle (fig. 27) in such a way as to have a total of 20 on each side.

47. ANOTHER NUMBER TRIANGLE.—Write the nine non-repetitive digits in the circles of the same triangle (fig. 27), but this time the sum on each side must be 17.

48. A MAGIC STAR.—The six-pointed star (fig. 28) is a "magic" one—the total in every row is the same:

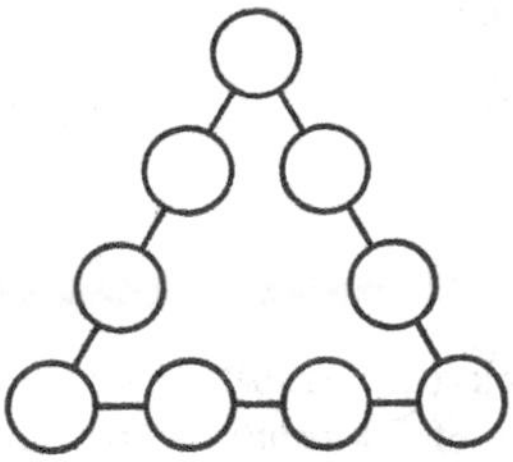

Fig. 27. Write the digits in the circles.

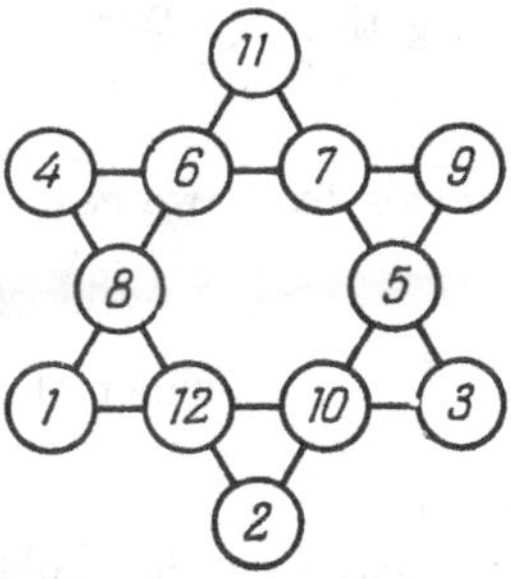

Fig. 28. A magic star.

$$4+6+7+9=26 \qquad 11+6+8+1=26$$
$$4+8+12+2=26 \qquad 11+7+5+3=26$$
$$9+5+10+2=26 \qquad 1+12+10+3=26$$

The sum of the numbers at the points, however, is different:

$$4+11+9+3+2+1=30$$

Can you perfect the star by placing the numbers in the circles in such a way as to make their sum in every row and at the points read 26?

Answers 37 to 48

37. All the three problems are insoluble. The magician and I could well afford to promise any prize money for their solution. To prove that, let us turn to algebra and analyze all three of them.

Payment of 5 rubles. Let us suppose that it is possible and that for this it will be necessary to have x number of 50-kopek coins, y number of 20-kopek coins and z number of 5-kopek coins. We now have the following equation:

$$50x+20y+5z=500 \text{ (or 5 rubles)}$$

Simplifying this by 5 we obtain:

$$10x+4y+z=100$$

Moreover, according to the problem, the total number of coins is 20; therefore, we have another equation:

$$x+y+z=20$$

Subtracting this equation from the first we have:

$$9x+3y=80$$

Dividing this by 3, we get:

$$3x+y=26\ {}^{2}/_{3}$$

But $3x$, i.e., the number of 50-kopek coins multiplied by 3, is, of course, an integer. So is y, the number of 20-kopek coins. The sum of these two numbers cannot be a fractional number. Therefore, it is nonsense to presume that the problem can be solved. It is insoluble.

In the same manner the reader may convince himself that the "reduced payment" problems are likewise insoluble. In the first case (3 rubles) we get the following equation:

$$3x+y=13\ {}^{1}/_{3}$$

And in the second (2 rubles):

$$3x+y=6\ {}^{2}/_{3}$$

Both, as you see, are fractional numbers.

So, the magician risked absolutely nothing by offering a big money prize for the solution of these problems. He would never have to pay for it.

It would have been another case if one had to give, say, 4 rubles—and not 5, 3 or 2—in 20 coins. It would have been easy to solve the problem then, and in seven different ways at that.*

38. $888+88+8+8+8=1,000$

39. Here are two solutions:

$$22+2=24;\quad 3^3-3=24$$

40. There are three solutions:

$$6\times6-6=30;\quad 3^3+3=30;\quad 33-3=30$$

41. The missing digits are restored gradually when we use the following method.

For convenience's sake let us number each line:

```
   ×1×  . . . . . . . . . . . . I
   3×2  . . . . . . . . . . . . II
   ———
   ×3×  . . . . . . . . . . . . III
 3×2×   . . . . . . . . . . . . IV
```

* Here is one of possible solutions: six 50-kopek coins, two 20-kopek coins and twelve 5-kopek coins.

$$
\begin{array}{rl}
\times 2 \times 5 & \ldots\ldots\ldots\ldots \text{V} \\
\hline
1 \times 8 \times 30 & \ldots\ldots\ldots\ldots \text{VI}
\end{array}
$$

It is easy to guess that the last digit in line III is 0; that is clear from the fact that 0 is at the end of line VI.

We next determine the meaning of the last × in line I: it is a digit that gives a number ending with 0 if multiplied by 2 and with 5 if multiplied by 3 (the number in line V ends with 5). There is only one digit to do that: 5.

It is not difficult to guess what hides behind x in line II: 8, for it is only this digit multiplied by 15 that gives a number ending with 20 (line IV).

Finally, it becomes clear that the first x in line I is 4, for only 4 multiplied by 8 gives a number that begins with 3 (line IV).

After that there will be no difficulty in restoring the remaining unknown digits: it will suffice to multiply the two factors which we have fully determined.

In the end, we have the following example of multiplication:

$$
\begin{array}{r}
415 \\
382 \\
\hline
830 \\
3320 \\
1245 \\
\hline
158530
\end{array}
$$

42. The same method applies to the solution of this problem. We get:

$$
\begin{array}{r}
325 \\
147 \\
\hline
2275 \\
1300 \\
325 \\
\hline
47775
\end{array}
$$

43. Here is the problem with all the digits restored:

$$
\begin{array}{l|l}
52650 & 325 \\
\cline{2-2}
325 & 162 \\
\cline{1-1}
2015 & \\
1950 & \\
\cline{1-1}
650 & \\
650 &
\end{array}
$$

44. To solve this problem we must know the rule governing the divisibility of a number by 11. A number is divisible by 11 if the difference between the sums of the odd digits and the even digits, counting from the right, is divisible by 11 or equal to 0.

For example, let us try 23,658,904.

The sum of the even digits:

$$3+5+9+4=21$$

And the sum of the odd digits is

$$2+6+8+0=16$$

The difference (subtracting the smaller number from the bigger, of course): 21—16=5 is not divisible by 11. Thus, the number is not divisible by 11 either.

Let us try another number, say, 7,344,535:

$$3+4+3=10$$
$$7+4+5+5=21$$
$$21-10=11$$

Since 11 is divisible by 11, the whole number is divisible too.

Now it is easy to guess in what order we must place our nine digits to get a number that is divisible by 11.

Here is an example:

352,049,786.

Let us verify it:

$$3+2+4+7+6=22$$
$$5+0+9+8=22$$

The difference (22-22) is 0. The number we have taken is, therefore, divisible by 11.

The biggest of such numbers is:

987,652,413

The smallest:

102,347,586

45. A patient reader can find nine examples of this sort. Here they are:

12×483 =5,796	27×198 =5,346	28×157 =4,396
42×138 =5,796	39×186 =7,254	4×1,738=6,952
18×297 =5,346	48×159 =7,632	4×1,963=7,852

46 and **47**. The solutions are shown in figures 29 and 30. The digits in the middle of each row may be transposed to get other solutions.

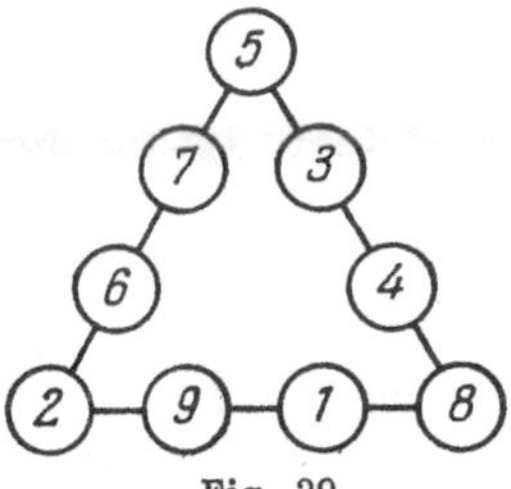

Fig. 29.

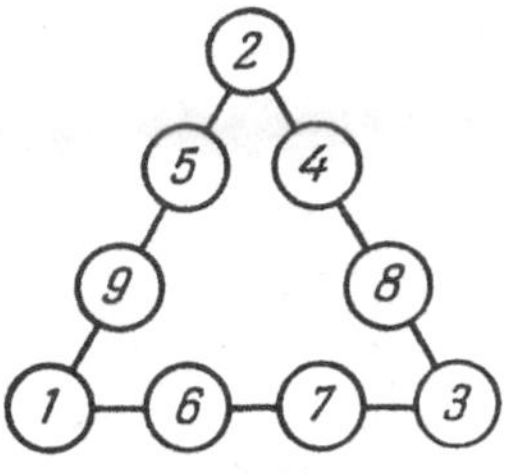

Fig. 30.

48. To see how the numbers are to be placed, let us proceed from the following assumption:

The sum of the numbers at the points is 26, while the total of all the numbers of the star is 78. Therefore, the sum of the numbers of the inner hexagon is 78—26=52.

Let us then examine one of the big triangles. The sum of the numbers on each of its sides is 26. If we add up the three sides we get 26×3=78. But in this case, the numbers at the points will each be counted twice. Since the sum of the numbers of the three inner pairs (i.e., of the inner hexagon) must, as we know, be 52, then the doubled sum at the points of each triangle is 78—52=26, or 13 for each triangle.

Our search now narrows down. We know, for instance, that neither 12 nor 11 can occupy the circles at the points. Then we can try 10 and

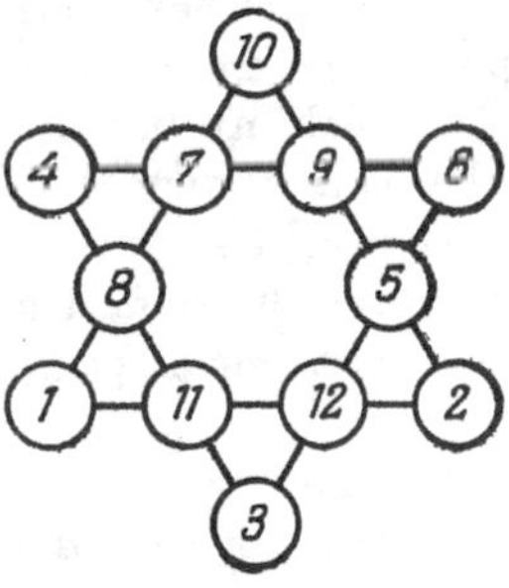

Fig. 31.

immediately come to the conclusion that the other two digits must be 1 and 2.

Now all we have to do is follow up and eventually we shall discover the arrangement we are seeking. It is shown in fig. 31.

CHAPTER VI

NUMBER GIANTS

49. A PROFITABLE DEAL.—We don't know when or where this took place. Perhaps it never did. That's even more probable. But whether fact or fable, it is an interesting story and well worth hearing (or reading).

I

A millionaire returned home extremely happy: he had met a person and the meeting, he said, promised to be most profitable.

"What luck!" he told his family. "It's true what people say: the rich have all the luck. At least I seem to have quite a bit of it. And it all happened quite unexpectedly. On my way home I met an inconspicuous person and probably I'd not have noticed him. But he learned that I was rich and approached me with a proposition. And that proposition, let me tell you, took my breath away.

" 'Let's make a deal,' he said. 'Every day for a month I'll bring you 100,000 rubles. Of course, I'll want something in return, but very little.' On the first day, he said, I'd have to pay him—it's too ridiculous to be true—just one kopek. I couldn't believe my ears.

" 'Just one kopek?' I asked him.

" 'Just one,' he confirmed. 'For the second 100,000 rubles you must pay me 2 kopeks.'

" 'And then?' I asked impatiently. 'What then?'

"'Well, for the third 100,000 rubles you pay me 4 kopeks, for the fourth 8 kopeks, for the fifth 16 kopeks. And so every day you must pay me double of what you paid on the previous day.'

"'And then what?'

Fig. 32. "Just one kopek.... "

"'Nothing. That's all. I won't ask for any more. Only you must abide by the agreement. Every day I'll bring you 100,000 rubles and every day you must pay me the sum we've agreed upon. The only stipulation is that you don't give up before the month is over.'

"Just think! He's giving away hundreds of thousands of rubles for a few kopeks. He's either a counterfeiter or a madman. Whatever he is, the deal is profitable. Can't afford to miss it.

"'All right,' I told him. 'Bring your money. I'll pay you what you're asking for. Only don't cheat me, don't bring any counterfeit notes.'

"'Don't worry,' he answered. 'You may expect me tomorrow morning.'

"Only I'm afraid he won't come. He's probably realized that he's done a silly thing. We'll see. Tomorrow isn't far away."

II

Early next morning there was a knock at the window. It was the stranger.

"Got your kopek ready?" he asked. "I've brought the money I promised."

True enough. The moment he came in, he took out a bundle of money, counted out 100,000 rubles—real at that—and said:

"Here's the sum as we have agreed. Now give me my kopek."

The millionaire put a copper coin on the table, his heart in his mouth, lest the stranger should change his mind and demand his money back. The visitor took the coin, weighed it in his palm and put it into his bag.

"I'll be here tomorrow at the same hour. Don't forget to have two kopeks ready."

The rich man couldn't believe his good luck: 100,000 rubles right from the moon! He counted the money, convinced himself that it was all there, without any counterfeit notes. After that he put it away, happily anticipating the next day.

In the night he started worrying. What if the stranger was a bandit in disguise and had only come to find out where he, the millionaire, kept his wealth, to rob him later on?

The rich man got up, bolted the doors more securely, repeatedly looked out of the windows, jumped up nervously every time he heard some noise and for a long time could not fall asleep. In the morning there was a knock at the window: the stranger was back. He counted off another 100,000 rubles, took the two kopeks promised him, put them in his bag and went off, saying:

Fig. 33. There was a knock at the window.

"Don't forget to have four kopeks ready tomorrow."

The rich man was happy beyond words—another 100,000 rubles in his pocket! And this time the visitor did not look like a bandit. In fact the millionaire no longer thought he was suspicious-looking. All he wanted was his few kopeks. What a crank! Should there be more of them in this world, then clever people would always live well. . . .

The stranger was on the dot on the third day, too, and the millionaire got his third 100,000 rubles, this time for 4 kopeks.

Another day, another 100,000 rubles—for 8 kopeks.

For the fifth 100,000 rubles the rich man paid 16 kopeks, and 32 kopeks for the sixth.

In the first seven days the millionaire received 700,000 rubles, paying a mere pittance for them:

$$1+2+4+8+16+32+64=1 \text{ ruble } 27 \text{ kopeks.}$$

The greedy man found this very much to his liking an he only thing he was sorry about was that the agreement was for only one month. That meant he would receive only 3,000,000 rubles. Shouldn't he try to talk the stranger into prolonging the agreement? No, better not. The man might realize that he was giving money away for nothing.

Meanwhile, the stranger continued to come every morning with his 100,000 rubles. On the eighth day he received 1 ruble 28 kopeks, on the ninth—2.56, on the tenth—5.12, on the eleventh—10.24, on the twelfth—20.48, on the thirteenth—40.96 and on the fourteenth —81.92.

The rich man paid readily. Hadn't he already received 1,400,000 rubles for something like 150 rubles?

But his joy was short-lived: he soon saw that the deal was not so profitable as it seemed. After 15 days he already had to pay hundreds of rubles, and not kopeks, and the payment sums were fast increasing. In fact this is what he paid:

For the fifteenth	100,000 rubles . .	163.84
For the sixteenth	100,000 rubles . .	327.68
For the seventeenth	100,000 rubles . .	655.36
For the eighteenth	100,000 rubles . .	1,310.72
For the nineteenth	100,000 rubles . .	2,621.44

Still, he was not losing yet. True, he had paid more than 5,000 rubles, but then hadn't he received 1,800,000 rubles in return?

The profit, however, was daily decreasing—and by leaps and bounds. Here is what the rich man paid after that:

For the twentieth 100,000 rubles . . .	5,242.88
For the twenty-first 100,000 rubles . . .	10,485.76
For the twenty-second 100,000 rubles . .	20,971.52
For the twenty-third 100,000 rubles . .	41,943.04
For the twenty-fourth 100,000 rubles . .	83,886.08
For the twenty-fifth 100,000 rubles . .	167,772.16
For the twenty-sixth 100,000 rubles . .	335,544.32
For the twenty-seventh 100,000 rubles .	671,088.64

Now he was paying very much more than receiving. It was time he should stop, but he could not violate the agreement.

And things went from bad to worse. All too late did the millionaire realize that the stranger had cruelly outwitted him, and that he would pay far more than he received.

On the 28th day the rich man had to pay over a million and the last two payments ruined him. They were astronomic:

For the twenty-eighth 100,000 rubles .	1,342,177.28
For the twenty-ninth 100,000 rubles . .	2,684,354.56
For the thirtieth 100,000 rubles . . .	5,368,709.12

When the visitor left for the last time, the millionaire sat down to count how much he had paid for the 3,000,000 rubles. The result was: 10,737,418 rubles 23 kopeks.

Only a little short of 11 million rubles! And it had all started with one kopek. The stranger would not have lost a kopek even if he had given him 300,000 rubles a day.

III

Before I finish with this story I shall show you a faster way of computing the millionaire's loss, i.e., a faster way of adding up the payments:

$$1+2+4+8+16+32+64, \text{ etc.}$$

It is not difficult to notice the following property of these numbers:

$$1=1$$
$$2=1+1$$
$$4=(1+2)+1$$
$$8=(1+2+4)+1$$
$$16=(1+2+4+8)+1$$
$$32=(1+2+4+8+16)+1, \text{ etc.}$$

We see that each number is equal to the sum of the preceding ones plus 1. Therefore, when we have to add all the numbers, for instance, from 1 to 32,768, we add to the last number (32,768) the sum of all the preceding ones or, in other words, the same number minus 1(32,768 —1) The result is 65,535.

Working by this method we can find out how much the millionaire has paid as soon as we know what he handed over the last time. His last payment is 5,368,709 rubles 12 kopeks. Thus, adding 5,368,709.12 and 5,368,709.11, we get the result we are seeking: 10,737,418.23.

Fig. 34. An interesting bit of news.

50. RUMOURS.—It is astonishing indeed how fast rumour spreads. Sometimes an incident or accident witnessed by just a few persons becomes the talk of the town within less than two hours. This extraordinary speed is more than astonishing, it's puzzling.

And yet, if you consider the whole thing arithmetically, you will see that there is really nothing wonderful about it—the thing becomes clear as day.

Let us analyze the following case.

I

A man living in the capital comes to a town with about 50,000 inhabitants and brings an interesting bit of news. At the house where he has stopped, he tells it to just three persons. This takes up, say, 15 minutes.

And so 15 minutes after the man has arrived, say at 8.15 a.m., the news is known to just four persons: himself and three local residents.

Each of the three hastens to tell it to three others. That takes another 15 minutes. In other words, half an hour later the news is the common knowledge of $4+(3\times3)=13$ persons.

In their turn, each of the nine persons who have learned the news last pass it on to three friends. By 8.45 a.m. the news is known to

$$13+(3\times9)=40 \text{ residents.}$$

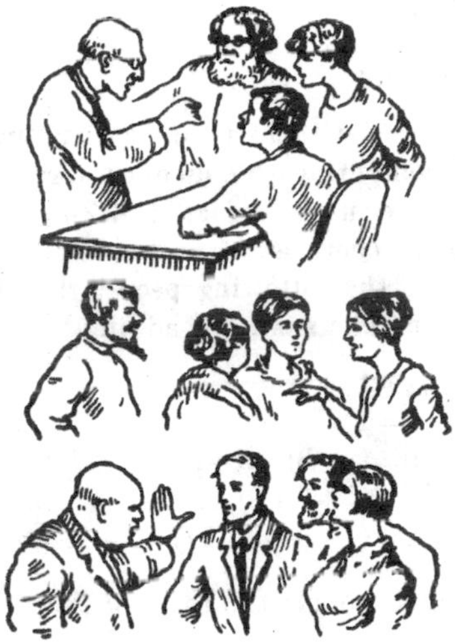

Fig. 35. Each tells the news to three others.

Fig. 36. The whole town will know the news by 10. 30.

If the rumour continues to spread in the same manner, i.e., if everyone who hears it passes it on to three others within the next 15 minutes, the result will be as follows:

By 9 a.m. the news will be known to $40+(3\times27)=121$ persons
By 9.15 a.m. „ „ „ „ „ $121+(3\times81)=364$ persons
By 9.30 a.m. „ „ „ „ „ $364+(3\times243)=1{,}093$ persons

In other words, within one and a half hours the news will be known to almost 1,100 persons. That does not seem too much for a town with a population of 50,000. In fact, some may think it will take quite a long time before the whole town knows it. Let us see how fast it will continue to spread:

By 9.45 a.m. the news will be known to $1{,}093+(3\times 729)=3{,}280$ persons
By 10 a.m. the news will be known to $3{,}280+(3\times 2{,}187)=9{,}841$ persons

In the next 15 minutes it will be the property of more than half of the town's population:

$$9{,}841+(3\times 6{,}561)=29{,}524 \text{ persons}$$

And this means that before it is 10.30 a.m. the news that only one man knew at 8 a.m. will be known to the entire town.

II

Let us see now how that is calculated. The whole thing boils down to the addition of the following numbers:

$$1+3+(3\times 3)+(3\times 3\times 3)+(3\times 3\times 3\times 3), \text{ etc.}$$

Fig. 37. How rumour spreads.

Perhaps there is an easier way of computing this number, like the one we used before $(1+2+4+8,$ etc.)? There is, if we take into account the following peculiarity of the numbers we are adding:

$$1=1$$
$$3=1\times 2+1$$
$$9=(1+3)\times 2+1$$
$$27=(1+3+9)\times 2+1$$
$$81=(1+3+9+27)\times 2+1, \text{ etc.}$$

In other words, each number is equal to double the total of the preceding plus 1.

Hence, to find the sum of all our numbers, from 1 to any number, it is enough to add to this last number half of itself (minus 1). For instance, the sum total of

$$1+3+9+27+81+243+729$$

equals $729+$half of 728, i.e., $729+364=1{,}093$

III

In our case, each resident passes the news to only three others. But if the residents of the town were more talkative and shared it not with three, but with five or even ten, the rumour would spread much faster. In the case of five, the picture would be as follows:

At 8 a.m. the news is known to		1 person
By 8.15 a. m.	1 + 5	6 persons
By 8.30 a.m.	6+(5 ×5)	31 „
By 8.45 a.m.	31+(25 ×5)	156 „
By 9 a.m.	156+(125 ×5)	781 „
By 9.15 a.m.	781+(625 ×5)	3,906 „
By 9.30 a.m.	3,906(3,125 ×5)	19,531 „

In short, it would be known to every one of the 50,000 residents before 9.45 a. m.

It would spread a lot faster if each man shared the news with ten others. Here we would get these very fast-growing numbers:

At 8 a.m. the news would be known to		1 person
By 8.15 a.m. . . .	1+10	11 persons
By 8.30 a.m. . . .	11+100	111 „
By 8.45 a.m. . . .	111+1,000	1,111 „
By 9 a.m	1,111+10,000	11,111 „

The next number is evidently 111,111, and that shows that the whole town would have heard the news shortly after 9 a.m. The news, in this case, would have taken a little over an hour to spread throughout the town.

51. **THE BICYCLE SWINDLE.**—In pre-revolutionary Russia there were firms which resorted to an ingenious way of disposing of average-quality goods. The whole thing would begin with an ad something like the following in popular newspapers and magazines:

A BICYCLE FOR 10 RUBLES!

You can get a bicycle for only 10 rubles
Take advantage of this rare chance

10 RUBLES INSTEAD OF 50

CONDITIONS SUPPLIED FREE ON APPLICATION

There were many, of course, who fell for the bait and wrote for the conditions. In return they would receive a detailed catalogue.

What the person got for his 10 rubles was not a bicycle, but four coupons which he was told to sell to his friends at 10 rubles each. The 40 rubles he thus collected he remitted to the company which then sent him the bicycle. And so, the man really paid only 10 rubles. The other 40 came from the pockets of his friends. True, apart from paying these 10 rubles, the purchaser had to go through quite a bit of trouble finding people who would buy the other four coupons, but then that did not cost him anything.

What were these coupons? What advantages did the purchaser get for his 10 rubles? He bought himself the right of exchanging this coupon for five similar coupons; in other words, he paid for the opportunity of collecting 50 rubles to purchase a bicycle which, in reality, cost him only 10 rubles, the sum he paid for the coupon. The new possessors of the coupons, in their turn, received five coupons each for further distribution, etc.

At the first glance, there was nothing fraudulent in the whole affair. The advertiser kept his promise: the bicycle really cost its purchaser only 10 rubles. Nor was the firm losing any money—it got the full price for its goods.

And yet, the thing was an obvious swindle. For this "avalanche," as it was called in Russia, caused losses to a great many people who were unable to sell the coupons they had purchased. It was these people who paid the firm the difference. Sooner or later, there came a moment when coupon-holders found it impossible to dispose of the coupons. That this was bound to happen you can see if you arm yourself with a pencil and a sheet of paper and calculate how fast the number of coupon-holders increased.

The first group of purchasers, receiving their coupons direct from the firm, usually had no difficulty in finding other buyers. Each member of this group drew four new participants into the deal.

The latter had to dispose of their coupons to 20 others (4×5) and to do that they had to convince them of the advantages of the purchase. Let us suppose that they were successful and that another 20 new participants were recruited.

The avalanche gathered momentum, and the 20 new holders of coupons had to distribute them among $20\times5=100$ others.

So far each of the original holders had drawn $1+4+20+100=125$ others into the game, and of these 25 received bicycles and the other 100 were given the hope of getting one—a hope for which they paid 10 rubles each.

The avalanche now smashed out of a narrow circle of friends and spread throughout the town where, however, it became increasingly hard to find new customers. The last 100 purchasers had to sell their coupons to 500 new victims who, in their turn, had to recruit another 2,500. The town was being flooded with coupons and it was becoming a difficult thing indeed to find people willing to buy them.

You will see that the number of the people drawn into the "bargain" increases along rumour-spreading lines (see above). Here is the pyramid of numbers we get:

1
4
20
100
500
2,500
12,500
62,500

If the town is big and the number of bicycle-riding people is 62,500, then the avalanche should peter out in the 8th round. By that time every person will have been drawn into the scheme. But only one-fifth will get bicycles, the rest will be in possession of coupons which they have no earthly chance to dispose of.

In a town with a bigger population, even in a modern capital with millions of people, the end comes only a few rounds later, because the pyramid of numbers grows with incredible speed. Here are the figures from the ninth round up:

312,500
1,502,500
7,812,500
39,062,500

In the 12th round, as you see, the scheme will have inveigled the population of a whole country, and 4/5 will have been swindled by the perpetrators of the fraud.

Let us see what they gain. They compel four-fifths of the population to pay for the goods bought by the remaining fifth, i.e., the former become the benefactors of the latter. Moreover, they get a whole army of volunteer salesmen—and zealous salesmen at that. A Russian writer justly called the affair "the avalanche of mutual fraud." And all that can be said of the thing is that people, who do not know how to calculate to guard themselves against frauds, are usually the ones who suffer.

52. REWARD.—Here is what, legend says, happened in ancient Rome.*

I

The Roman general Terentius returned home from a victorious campaign with trophies and asked for an audience with the emperor.

* This is a liberal translation from a Latin manuscript in the keeping of a private library in Britain.

The latter received him very kindly, thanked him for what he had done for the empire and promised him a place in the Senate that would befit his dignity.

But that was not the reward Terentius wanted.

"I have won many a victory to enhance thy might and glorify thy name," he said. "I have not been afraid of death, and had I more lives than one I would have willingly sacrificed them for thee. But I am tired of fighting. I am no longer young and the blood in my veins is no longer hot. It is time I retired to the home of my ancestors and enjoyed life."

"What wouldst thou like then, Terentius?" the emperor asked.

"I pray thy indulgence, O Caesar! I have been a warrior almost all my life, I have stained my sword with blood, but I have had no time to build up a fortune. I am a poor man. . . ."

"Continue, brave Terentius," the emperor interrupted him.

"If thou wouldst reward thy servant," the encouraged general went on, "let thy generosity help me to live my last days in peace and plenty. I do not seek honours or a high position in the almighty Senate. I should like to retire from power and society to rest in peace. O Caesar, give me enough money to live the rest of my days in comfort."

The emperor, the legend says, was not a generous man. He was a miser, in fact, and it hurt him to part with money. He thought for a moment before answering the general.

"What is the sum thou wouldst consider adequate?" he finally asked.

"A million denarii, O Caesar."

The emperor again fell silent. The general waited, his head low.

"Valiant Terentius," the emperor said at last. "Thou art a great general and thy glorious deeds indeed deserve a worthy reward. I shall give thee riches. Thou wilt hear my decision at noon tomorrow."

Terentius bowed and left.

II

The next day Terentius returned to the palace.

"Hail, O brave Terentius!" the emperor said.

The general bowed reverently.

"I have come, O Caesar, to hear thy decision. Thou hast graciously promised to reward me."

"Yes," the emperor answered, "I would not want a noble warrior like thee to receive a niggardly reward. Hark to me. In my treasury there are 5 million brass coins worth a million denarii. Now listen carefully. Thou wilt go to my treasury, take one coin and bring it here. On the next day thou wilt go to the treasury again and take another coin worth twice the first and place it beside the first. On the third day thou wilt get a coin worth four times the first, on the fourth day eight times, on the fifth sixteen times, and so on. I shall order to have coins of the required value minted for thee every day. And so long as thou hast the strength, thou mayst take the coins out of my treasury. But thou must do it thyself, without any help. And when thou canst no longer lift the coin, stop. Our agreement will have ended then, but all the coins thou wilt have taken out will be thy reward."

Terentius listened greedily to the emperor. In his imagination he saw the huge number of coins he would take out of the treasury.

"I am thankful for thy generosity, O Caesar," he answered happily. "Thy reward is wonderful indeed!"

Fig. 38. The first coin.

Fig. 39. The seventh coin.

Fig. 40. The ninth coin.

III

And so Terentius began his daily pilgrimages to the treasury near the emperor's audience hall, and it was not difficult to bring the first coins there.

On the first day Terentius took a small coin that was 21 mm. in diameter and weighed 5 grammes.

Carrying his second, third, fourth, fifth and sixth coins was quite easy too, for all they weighed was 10, 20, 40, 80 and 160 grammes.

The seventh coin weighed 320 grammes and was $8^{1}/_{2}$ cm. (or, to be exact, 84 mm.*) in diameter.

On the eighth day Terentius had to take a coin worth 128 original coins. It weighed 640 grammes and was about $10^{1}/_{2}$ cm. in diameter.

Fig. 41. The eleventh coin.

Fig. 42. The thirteenth coin.

Fig. 43. The fifteenth coin.

On the ninth day he brought to the emperor a coin worth 256 times the first coin, weighing more than 1.250 kilogrammes and 13 cm. in diameter.

On the twelfth day the coin was almost 27 cm. in diameter and weighed 10.250 kilogrammes.

The emperor, who had always greeted him graciously, found it hard to conceal his triumph. He saw that Terentius had been 12 times to the treasury and had brought back only a little over 2,000 brass coins.

The thirteenth day gave Terentius a coin worth 4,096 original coins. It was about 34 cm. in diameter and weighed 20,5 kilogrammes. The

* If the coin is 64 times heavier than the ordinary one, it is only four times greater in diameter and thickness, because $4 \times 4 \times 4 = 64$. This should be remembered as we calculate the size of the coins later on in the story.

next day the coin was still heavier and bigger: 41 kilogrammes in weight and 42 cm. in diameter.

Fig. 44. The sixteenth coin.

Fig. 45. The seventeenth coin.

"Art thou not tired, my brave Terentius?" the emperor asked, hardly able to abstain from smiling.

"No, Caesar," the general answered, frowning and wiping the sweat off his brow.

Then came the fifteenth day. The burden was heavier than ever and Terentius made his way slowly to the audience room, carrying a coin that was worth 16,384 original coins. It was 53 cm. in diameter and weighed almost 82 kilogrammes—the weight of a tall warrior.

On the sixteenth day the general's legs shook as he carried the burden on his back. It was a coin worth 32,768 original coins and weighing 164 kilogrammes. Its diameter was 67 cm.

Terentius came to the audience hall breathing hard and looking very tired. The emperor met him with a smile. . . .

When the general returned there on the following day, he was greeted with laughter. He could no longer carry the coin and had to roll it in. It was 84 cm. in diameter, weighed 328 kilogrammes and was worth 65,536 original coins.

The eighteenth day was the last day he could enrich himself. His visits to the treasury and thence to the audience hall came to an end. This time he had to bring a coin worth 131,072 original coins, more than a metre in diameter and weighing 655 kilogrammes. Using his spear as a steering lever he rolled the coin in. It fell with a thud at the emperor's feet.

Terentius was completely exhausted.

"Enough. . ." he gasped.

Fig. 46. The eighteenth coin.

The emperor could hardly restrain himself from laughing with delight. He had outwitted the general. Later he ordered the treasurer to calculate how much Terentius had taken out of the treasury.

The treasurer did so.

"Thanks to thy generosity, O Caesar, the valiant Terentius hath received 262,143 brass coins as a reward."

And so, the stingy emperor paid the general something like one-twentieth of the million denarii the latter had asked for.

* * *

Let us check on the treasurer and, at the same time, the weight of the coins. What Terentius took out of the treasury was:

	The equivalent of	
On the first day	1 coin weighing	5 gr.
On the second day	2 coins weighing	10 gr.
On the third day	4 coins weighing	20 gr.
On the fourth day	8 coins weighing	40 gr.
On the fifth day	16 coins weighing	80 gr.
On the sixth day	32 coins weighing	160 gr.
On the seventh day	64 coins weighing	320 gr.
On the eighth day	128 coins weighing	640 gr.
On the ninth day	256 coins weighing	1.280 kg.
On the tenth day	512 coins weighing	2.560 kg.
On the 11th day	1,024 coins weighing	5.120 kg.
On the 12th day	2,048 coins weighing	10.240 kg.
On the 13th day	4,096 coins weighing	20.480 kg.

On the 14th day	8,192 coins weighing	40.960 kg.
On the 15th day	16,384 coins weighing	81.920 kg.
On the 16th day	32,768 coins weighing	163.840 kg.
On the 17th day	65,536 coins weighing	327.680 kg.
On the 18th day	131,072 coins weighing	655.360 kg.

We already know how simple it is to calculate the sum of the numbers of the second column (the same rule as the one applied on page 67). In this case it is 262,143. While Terentius asked for 1,000,000 denarii, i.e., 5,000,000 brass coins. Therefore, he received:

$$5{,}000{,}000 : 262{,}143 = 19 \text{ times less.}$$

53. THE LEGEND ABOUT CHESS.*—Chess is one of the oldest games in the world. It was invented many, many centuries ago and it is not surprising, therefore, that there are so many legends about it--legends that it is, of course, impossible to verify. I should like to relate one of them. It is not necessary to know how to play chess to understand the legend: it is enough to know that it is played on a checkered board with 64 squares.

I

Chess, legend has it, comes from India.

King Sheram was thrilled by the huge number of clever moves one could make in the game. Learning that its author was one of his subjects, he commanded the man to be brought before him in order to reward him personally for his marvellous invention.

The inventor, a man called Sessa, appeared before the king—a simply clad scholar who made his living by teaching.

"I wish to reward thee well for thy wonderful invention," the king greeted Sessa.

The sage bowed.

"I am rich enough," the king continued, "to satisfy thy most cherished wish. Just name what thou wouldst have and thou shalt have it."

Sessa was silent.

"Don't be shy," the king encouraged him. "Say what thou wouldst like to have. I shall spare nothing to satisfy thy wish."

"Thy kindness knows no bounds, O Sire," the scholar replied. "But give me time to consider my reply. Tomorrow, after I have well thought about it, I shall tell thee my request."

The next day Sessa surprised the king by his extremely modest request.

* This legend is my own adaptation.—*Author.*

Fig. 47. "Two for the second...."

"Sire," he said, "I should like to have one grain of wheat for the first square on the chessboard."

"A grain of ordinary wheat?" The king could hardly believe his ears.

"Yes, Sire. Two for the second, four for the third, eight for the fourth, 16 for the fifth, 32 for the sixth. . . ."

"Enough," the king was irritated. "Thou shalt get thy grains for all the 64 squares of the chessboard as thou wishest: every day double the amount of the preceding day. But know thou that thy request is not worthy of my generosity. By asking for such a trite reward thou hast shown disrespect for me. Truly, as a teacher thou couldst have shown a better example of respect for thy king's kindness. Go! My servants shall bring thee thy sack of grain."

Sessa smiled and went out, and then waited at the gate for his reward.

II

At dinner the king remembered Sessa and inquired whether the "foolhardy" inventor had been given his miserable reward.

"Sire," he was told, "thy command is being carried out. Thy sages are calculating the number of grains he is to receive."

The king frowned. He was not accustomed to seeing his commands fulfilled so slowly.

In the evening, before going to bed, the king again asked whether Sessa had been given his bag of grain.

"Sire," was the reply, "thy mathematicians are working incessantly and hope to compute the sum ere dawn breaks."

"Why are they so slow?" the king demanded angrily. "Before I awake Sessa must be paid in full, to the last grain. I do not command twice!"

In the morning, the king was told that the chief court mathematician had asked for an audience.

The king ordered him to be admitted.

"Before thou tellest me what thou hast come for," King Sheram began, "I want to know whether Sessa has been given the niggardly reward he asked for."

"It is because of this that I have dared come before thy eyes so early in the morning," the old sage replied. "We have worked conscientiously to calculate the number of grains Sessa wants. It is tremendous. . . ."

"However tremendous," the king interrupted him impatiently ,"my granaries can easily stand it. The reward has been promised and must be paid!"

"It is not within thy power, O Sire, to satisfy Sessa's wish. Thy granaries do not hold the amount of grain Sessa has asked for. There is not that much grain in the whole of thy kingdom; in fact, in the whole world. And if thou wouldst keep thy word, thou must order all the land in the world to be turned into wheat fields, all the seas and oceans drained, all the ice and snow in the distant northern deserts melted. And if all this land is sown to wheat, then perhaps there will be enough grain to give Sessa."

The king listened awe-struck to the wise man.

"Name this giant number," he said thoughtfully.

"It is 18,446,744,073,709,551,615, O Sire!" the sage replied.

III

So goes the legend. We do not know whether it was really so, but that the reward would run into such a number is not difficult to see: with a little patience we can calculate it ourselves. Starting with one we must add up the numbers: 1, 2, 4, 8, etc. The result of the 63rd power of 2 will show us how much the inventor was to receive for the 64th square. Following the pattern shown on p. 67 we shall easily find the number of grains if we find the value of 2^{64} and subtract 1. In other words, we must multiply 64 2's:

$$2\times2\times2\times2\times2\times2, \text{ etc., } 64 \text{ times.}$$

To facilitate calculation we shall divide these 64 factors into 6 groups of 10 2's, the last group to contain 4 2's. The product of 2^{10} is 1,024, and of 2^4 is 16. Hence, the value we seek is

$$1{,}024 \times 1{,}024 \times 1{,}024 \times 1{,}024 \times 1{,}024 \times 1{,}024 \times 16$$

Multiplying 1,024 by 1,024 we get 1,048,576.

What we have to find now is

$$1{,}048{,}576 \times 1{,}048{,}576 \times 1{,}048{,}576 \times 16$$

and subtract 1 from the result—and then we shall know the number of grains:

18,446,744,073,709,551,615

If you want to have a clear picture of what this giant number is really like, just imagine the size of the granary that will be required to store all this grain. It is well known that a cubic metre of wheat contains 15,000,000 grains. Hence, the reward asked by the inventor of chess would require a granary of approximately 12,000,000,000,000 cubic metres or 12,000 cubic kilometres. If we take a granary 4 metres in height and 10 metres in width, its length must be 300,000,000 kilometres, i. e., twice the distance from the Earth to the Sun.

The king was unable to satisfy Sessa's request. But had he been clever in mathematics, he would have easily avoided promising such a huge reward—all he should have done was to offer Sessa to count the grains himself—one by one.

Indeed, if Sessa had counted the grain day and night, without stopping, taking a second for each grain, he would have counted 86,400 grains on the first day. One million grains would have taken him no fewer than 10 days to count. He would have taken about six months to count the grains in one cubic metre of wheat—that would have given him 27 bushels. Counting without interruption for 10 years, he would have counted off about 550 bushels. You will see that even if Sessa had devoted all the remaining years of his life to counting the grain, he would have got only an insignificant part of the reward.

54. RAPID REPRODUCTION.—A ripe poppy is full of minute seeds, and from each a new plant may be grown. How many poppy plants would we have if all the seeds we planted grew into plants? To find this, we must know how many seeds there are in each poppy. A tedious job, perhaps, but the result is so interesting that it is well worth while to arm oneself with patience and do the job thoroughly. First, you will find that each poppy has on the average 3,000 seeds.

What next? You will observe that if there is enough arable land around our poppy plant, each seed will grow into a plant and that will give us 3,000 plants by the following summer. A whole poppy field from just one poppy.

Let us see what comes next. Each of these 3,000 plants will bring us at least one poppy (very often more) with 3,000 seeds. Grown into plants, they will each give us 3,000 new plants. Hence, at the end of the second year we shall have no fewer than

$$3{,}000 \times 3{,}000 = 9{,}000{,}000 \text{ plants}$$

It is easy to calculate that at the end of the third year the progeny of our single poppy will be

$$9{,}000{,}000 \times 3{,}000 = 27{,}000{,}000{,}000$$

And at the end of the fourth:

$$27{,}000{,}000{,}000 \times 3{,}000 = 81{,}000{,}000{,}000{,}000$$

At the end of the fifth year there will not be enough space on earth for our poppies, for the number will then be:

$$81{,}000{,}000{,}000{,}000 \times 3{,}000 = 243{,}000{,}000{,}000{,}000{,}000$$

And the entire surface of the earth, i. e., of all the continents and islands, is only 135,000,000 square kilometres or

$$135{,}000{,}000{,}000{,}000 \text{ square metres}$$

and that is approximately 2,000 times less than the number of poppy plants that will have grown by then.

You will see that if all the poppy seeds were to grow into plants, the progeny of one poppy would cover the entire land surface of the globe within five years—with 2,000 to a square metre. The little poppy seed does conceal a giant number, doesn't it?

We can try the same thing with some other plant that yields fewer seeds, and yet come to the same result—only its progeny would then take slightly longer than five years to cover the entire surface of the earth. For instance, take a dandelion that yields on the average 100 seeds a year.* If all these seeds grew into plants, we would have:

At the end of the first year	1 plant
At the end of the second year	100 plants
At the end of the third year	10,000 plants
At the end of the fourth year	1,000,000 plants
At the end of the fifth year	100,000,000 plants
At the end of the sixth year	10,000,000,000 plants
At the end of the seventh year	1,000,000,000,000 plants
At the end of the eighth year	100,000,000,000,000 plants
At the end of the ninth year	10,000,000,000,000,000 plants.

* There are dandelions that yield up to 200 seeds, though they are rare.

Fig. 48. A dandelion yields 100 seeds a year.

This is 70 times more than there are square metres on the entire land surface of the globe.

Therefore, at the end of the ninth year all the continents would be covered with dandelions—70 of them to a square metre.

Then how is it that this does not happen? The reason is simple: the overwhelming majority of seeds perish before they give root to new plants—they either fall on sterile soil, are stifled by other plants if they do take root or are destroyed by animals. If there were no mass destruction of seeds and plants, each of them would cover our planet in no time at all.

All this applies not only to plants, but to animals too. If they did not die, the earth would sooner or later be overcrowded with the progeny of just one pair of animals. The locust swarms that cover vast areas are graphic evidence of what would happen if death did not hinder the growth of living organisms. Within a score or so years, our continents would be covered with forests and steppes teeming with millions of animals fighting each other for living space. The oceans would have so much fish in them that navigation would be out of question, and we would hardly see daylight for the multitude of birds and flies swarming in the air.

Let us take, for instance, the common fly which is appallingly prolific. Suppose that each female fly lays 120 eggs and that in the course of the summer out of these 120 eggs 7 generations of flies will hatch—half of them female. Let us suppose that the first eggs are laid on April 15 and that the female flies hatched grow sufficiently within 20 days to lay eggs themselves. The picture will be as follows:

On April 15, a female fly lays 120 eggs; at the beginning of May there will hatch 120 flies, 60 of them female.

On May 5, each female lays 120 eggs and in the middle of the month there will hatch $60 \times 120 = 7{,}200$ flies, 3,600 of them female.

On May 25, each of these 3,600 female flies will lay 120 eggs and at the beginning of June there will be $3,600\times120=432,000$ flies, 216,000 of them female.

On June 14, each of the female flies will lay 120 eggs and at the end of the month there will hatch 25,920,000 flies, including 12,960,000 female flies.

On July 5, 12,960,000 female flies will lay 120 eggs each that will bring 1,555,200,000 flies (777,600,000 female flies).

On July 25, there will be 93,312,000,000 flies, 46,656,000,000 of them female flies.

On August 13, the number will be 5,598,720,000,000, of which 2,799,360,000,000 will be female flies.

On September 1, there will hatch 355,923,200,000,000 flies.

To get a clearer picture of this huge mass of flies that can hatch in just one summer if nothing were done about it and if none were to die, let us see what happens if they form a line. A fly is 5 mm. long and this line would be 2,500,000,000 kilometres long, i. e., 18 times longer than the distance from the Earth to the Sun (or approximately the distance from the Earth to Uranus, one of remote planets).

Dig. 49. They would stretch from the Earth to Uranus.

In conclusion it would be well to cite some *facts* on the extraordinarily rapid reproduction of animals in favourable conditions.

Originally, there were no *sparrows* in America. They were brought to the United States with the express aim of destroying pests. Sparrows, as you know, feed on voracious caterpillars and other vermin. The sparrows, it seems, liked the country—there were no beasts or birds of prey to destroy them, and they began to reproduce at a very fast pace. The number of pests grew less and less, but that of sparrows increased by leaps and bounds. Eventually there were not enough ants for them and they turned to destroying crops.* A regular war was declared on the sparrows and it proved so expensive that legislation was later passed, prohibiting the import to the United States of any animals.

And here is another example. There were no *rabbits* in Australia

* In Hawaii they forced all the other little birds out.

when it was discovered by the Europeans. The first rabbits were brought there at the end of the 18th century and since there are no beasts of prey that feed on rabbits, the latter began to multiply at an extremely fast rate. Soon hordes of rabbits were overrunning Australia and destroying crops. The calamity became nation-wide, and huge sums of money were spent on the extermination of these rodents. It was only the resolute measures taken by the people that put an end to the catastrophe. A more or less similar thing happened later in California.

The third story comes from Jamaica. There were a great many poisonous snakes there and to destroy them it was decided to bring the *secretary bird*, which is known as a rabid enemy of snakes. True, the

Fig. 50. The secretary bird.

number of snakes soon grew less, but then the number of field rats—which the snakes used to devour—began to grow. The rodents caused so much damage to sugar-cane plantations that the farmers had to declare a war of extermination on them and brought four pairs of Indian *mongooses*, which are known as enemies of rats. They were allowed freely to reproduce and soon the island was full of them. Within some ten years they exterminated almost all the rats, but in doing that they became omnivorous and began to attack puppies, kids, sucking-pigs and chickens, and to destroy eggs. As they grew in number, they swarmed into orchards, wheat fields and plantations. The inhabitants of the island turned against their former allies, but succeeded only partially in checking the damage.

55. A FREE DINNER.—Ten young men decided to celebrate their graduation from secondary school with a dinner at a restaurant.

When they got together and the first dish was served, they started arguing as to which seat each should occupy. Some of them proposed to do so alphabetically, others according to age, still others according to height. There was one who even suggested to do so according to graduation marks. The argument went on and on, the meal grew cold and still none would sit down. The problem was solved by the waiter.

Fig. 51. "Sit down where you are...."

"Listen, my young friends," he said, "stop arguing. Sit down where you are and hear what I have to say.

"Let one of you write down the order in which you are now sitting. Return here tomorrow and sit down in a different order. After tomorrow do it in yet a different way, and so on until you have tried all the combinations. And when the time again comes for you to sit down at the places where you are now sitting, I promise to serve you free of charge any delicacies you may like."

The suggestion was tempting and it was decided to meet at the restaurant every day and to try every possible way of sitting around the table so as to get the free dinners the waiter had promised.

That day, however, never came, and not because the waiter failed to keep his word, but because there were too many different ways for ten men to sit around the table—in fact, 3,628,800 of them. And to try all of them it would take almost 10,000 years—as you will see.

Perhaps you may not believe that there are so many ways for ten persons to sit around a table? To make it as simple as possible, let us start with three objects that we shall call A, B, and C.

What we want to find is how many ways there are of rearranging these objects. First, let us put C aside and do it with just two objects. We will see that there are only two ways of rearranging them.

Fig. 52. Only two ways of rearranging two things.

Now let us add C to each of these pairs. We can do it in three different ways:

1. We can put C *behind* the pair,
2. *Before* the pair, and
3. *Between* the two objects.

There are evidently no other ways of placing it. And since we have *two* pairs, AB and BA, we have

$$2\times3=6 \text{ ways of rearranging the objects.}$$

The ways these objects are rearranged are shown in fig. 53.

Let us proceed—with four objects: A, B, C, and D. For the time being we shall put D aside and do all the rearrangements with three objects. We already know that there are 6 ways of doing that. How

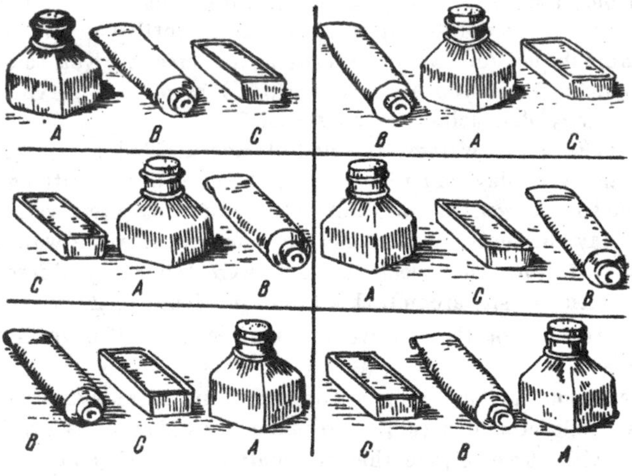

Fig. 53. Three things can be arranged in six ways.

many ways are there to add the fourth object (D) to each of the 6 arrangements of the other three objects? Let us see. We can

1. Put D *behind* the three objects,
2. *Before* them,
3. *Between* the first and second objects, and
4. *Between* the second and third objects.

Therefore, we have

$$6 \times 4 = 24 \text{ arrangements,}$$

and since $6 = 2 \times 3$ and $2 = 1 \times 2$, then the number of all the arrangements may be written as follows:

$$1 \times 2 \times 3 \times 4 = 24$$

Now if we apply the same method with five objects, we get the following:

$$1 \times 2 \times 3 \times 4 \times 5 = 120$$

And for six:

$$1 \times 2 \times 3 \times 4 \times 5 \times 6 = 720, \text{ etc.}$$

Let us now return to the ten young men. The number of possible arrangements in this case—if we take the trouble to calculate it—will be as follows:

$$1 \times 2 \times 3 \times 4 \times 5 \times 6 \times 7 \times 8 \times 9 \times 10$$

The result will be the number we mentioned above:

$$3{,}628{,}800$$

Calculation would have been far more complicated if half of the young people were girls, and they would want to sit with each young man in turn. Although the number of arrangements in this case would be much smaller, it would be harder to compute it.

Let one young man sit down wherever he wants at the table. The other four, leaving empty chairs between them for the girls, can sit down in $1 \times 2 \times 3 \times 4 = 24$ different ways. Since there are 10 chairs, the first young man can sit down in 10 different ways; therefore, there are $10 \times 24 = 240$ different ways in which the young men can occupy their seats around the table.

How many ways are there in which the five girls can occupy the empty seats between the young men? Obviously $1 \times 2 \times 3 \times 4 \times 5 = 120$ ways. Combining each of the 240 positions of each young man with each of the 120 positions of each girl, we come to the number of possible arrangements, which is:

$$240 \times 120 = 28{,}800$$

This, of course, is very much less than the 3,628,800 arrangements for the young men and would take slightly less than 79 years. And that means that the young people would get a free dinner from the heir of the waiter, if not from the waiter himself, by the time they were about 100 years old—provided they lived that long.

Now that we have learned how to calculate the number of arrangements, we can determine the number of combinations of blocks in a "Fifteen Puzzle"* box. In other words, we can compute the number of problems this game can set a player. It is easy to see that the task is to determine the total number in which the blocks can be rearranged. To do that, we know, we must effect the following multiplication:

$$1 \times 2 \times 3 \times 4 \times 5 \times 6 \times 7 \times 8 \times 9 \times 10 \times 11 \times 12 \times 13 \times 14 \times 15$$

The answer is

$$1{,}307{,}674{,}365{,}000$$

Half of this huge number of problems are insoluble. There are, therefore, more than 600,000,000,000 problems for which there are no solutions. The fact that people never even suspected that explains the craze for the "Fifteen Puzzle."

Let us also note that if it were possible to shift one block every second, it would take more than 40,000 years to try all the possible combinations, and that if one sat at it uninterruptedly.

As we come to the close of our talk about arrangements, let us solve a problem right out of school life.

Let us suppose there are 25 pupils in a class. How many ways are there to seat them?

Those who have well understood the problems we explained above will not find any difficulty in solving this one. All we have to do is to multiply the 25 numbers, thus:

$$1 \times 2 \times 3 \times 4 \times 5 \times 6 \ldots \times 23 \times 24 \times 25$$

Mathematics shows many ways of simplifying various operations, but there is none for the one mentioned above. The only way to do it *correctly* is to multiply all these numbers. And the only thing that will save time is an appropriate arrangement of multipliers. The result is stupendous—it runs into 26 digits—so stupendous that it is beyond our power of imagination.

---

* The square in the lower right-hand corner must always remain vacant.

Here it is:

15,511,210,043,330,985,984,000,000

Of all the numbers we have encountered so far this one, of course, is the biggest and therefore takes the palm as *the* number giant. Compared with it, the number of drops in all oceans and seas is quite modest.

56. A TRICK WITH COINS.—In my boyhood my brother, I recall, showed me an interesting game with coins. First he put three saucers in a line and then placed five coins of different denomination (one-ruble coin, 50-kopek coin, 20-kopek coin, 15-kopek coin and 10-kopek coin* in the first saucer, one atop the other in the order given. The task was to transpose these coins to the third saucer, observing the following three rules:

1) It is permitted to transpose only one coin at a time;

2) It is not permitted to place a bigger coin on a smaller one; and

3) It is permitted to use the middle saucer *temporarily*, observing the first two rules, but in the end the coins must be in the third saucer and in the original order.

Fig. 54. My brother showed me an interesting game.

"The rules," my brother said, "are quite simple, as you see. Now get to it."

I took the 10-kopek coin and put it into the third saucer, then I placed the 15-kopek into the middle saucer. And then I got stuck. Where was I to put the 20-kopek coin? It was bigger than both!

---

* The game can be played with any five coins of different size.

"Well?" my brother came to my assistance. "Put the 10-kopek coin on top of the 15 kopeks. Then you will have the third saucer for the 20-kopek coin."

I did that. But it did not mean the end of my difficulties. Where to put the 50-kopek coin? I soon saw the way out: I put the 10-kopek coin into the first saucer, the 15-kopek coin into the third and then transposed the 10-kopek coin there too. Now I could place the 50-kopek coin in the second saucer. Then, after numerous transpositions, I succeeded in moving the ruble coin from the first saucer and eventually had all the pile in the third.

"Well, how many moves did you make altogether?" my brother asked, praising me for the way I had solved the problem.

"Don't know. I didn't count."

"All right, let's count. It would be interesting to know how to get it done with the least possible number of moves. Let's suppose we had only two coins—15- and 10-kopek—and not five. How many moves would you require then?"

"Three—the 10-kopek coin would go into the middle saucer, the 15-kopek coin into the third and then the 10 kopeks over it."

"Correct. Let's add another coin—the 20-kopek—and see how many moves we need to transpose the pile. First we move the two smaller coins to the middle saucer. To do that, as we know it, we need three moves. Then we move the 20-kopek coin to the third saucer. That's another move. Then we move the two coins from the second saucer to the third and that's another three moves. Therefore, we have to do $3+1+3=7$ moves."

"Let me calculate the number of moves we would require for four coins," I interrupted him. "First, I move the three smaller coins to the middle saucer. That's seven moves. Then I transpose the 50-kopek coin to the third saucer. That's another move. And finally the three smaller coins to the third saucer and that's another seven moves. Altogether it will be $7+1+7=15$ moves."

"Excellent. And what about five coins?"

"Easy: $15+1+15=31$," I answered promptly.

"Well, I see you've caught on. But I'll show you a still easier way of doing it. Take the numbers we have obtained: 3, 7, 15 and 31. All of them represent 2 multiplied by itself once or several times, minus 1. Look."

And my brother wrote down the following table:

$$3=2\times2-1$$
$$7=2\times2\times2-1$$

$$15=2\times2\times2\times2-1$$
$$31=2\times2\times2\times2\times2-1$$

"I see it now. We multiply 2 by itself as many times as there are coins to be transposed and then subtract 1. Now I know how to calculate the number of moves for any pile of coins. For instance, if we have seven coins, the operation will look as follows:

$$2\times2\times2\times2\times2\times2\times2-1=128-1=127."$$

"Well, now you know this ancient game," my brother said. "There's only one other rule that you should bear in mind: if the number of coins is odd, then you put the first into the third saucer; if it's even, you start with the second saucer."

"Is the game really ancient? I thought it was your own!" I exclaimed.

"No, all I did was to modernize it with coins. The game's very, very old and comes from India, I think. There's a very interesting legend connected with it. In Benares there's a temple and it is said that when Brahma created the world he put up three diamond sticks there and around one of them he placed 64 gold rings, with the biggest at the bottom and the smallest on top. The priests had to work day and night without a stop, transposing the rings from one stick to another, using the third as an aid—the rules were the same as in the case of coins: they were allowed to transpose only one ring at a time and forbidden to place a bigger ring on top of a smaller one. When all the rings are transposed, the legend says, the world will come to an end."

"Then the world should have perished long ago, if one is to believe this legend."

"You think the transposition of these 64 rings doesn't take much time, do you?"

"Of course, it doesn't. Let's say it takes a second for each move. That means in an hour one can make 3,600 transpositions."

"Well?"

"That'll be about 100,000 a day and about 1,000,000 in ten days, and I'm sure you can transpose all of 1,000 rings with 1,000,000 moves."

"You're wrong there. To transpose these 64 rings you'll require neither more nor less than 500,000 million years!"

"But why? The total number of transpositions will be equal to 2 multiplied by itself 64 times minus 1, that is, to. . . . Wait, I'll tell you the result in a second."

"Fine. And while you're doing all this multiplication job I'll have enough time to attend to some business."

Fig. 55. The priests worked day and night.

My brother left and I busied myself with calculation. First I found the value of 2^{16} and then multiplied the result—65,536—by itself and then the result again by itself and subtracted 1. What I got after that was

18,446,744,073,709,551,615*

My brother was right, after all.

Incidentally, you might be interested in learning how old our Earth is. Well, scientists have worked that out—though only approximately:

The Sun has existed	10,000,000,000,000	years
The Earth	2,000,000,000	"
Life on Earth	300,000,000	"
Human beings	300,000	"

57. A BET.—We were having lunch at our holiday home when the talk turned to determination of the *probability* of a coincidence. One of the fellows, a young mathematician, took out a coin and said:

"Look, I'll toss this coin on the table without looking. What's the probability of a head-up turn?"

"You'd better explain what 'probability' is," the rest chorussed. "Not everyone knows what it is."

"Oh, that's simple. There are only two possible ways in which a

* We know this figure: it was the number of grains Sessa asked as a reward for inventing chess.

Fig. 56. Head or tail.

Fig. 57. A die.

coin may fall (fig 56): either head or tail. Of these only one will be a favourable occurrence. Thus we come to the following relation:

$$\frac{\text{The number of favourable occurrences}}{\text{The number of possible occurrences}} = \frac{1}{2}$$

"The fraction $\frac{1}{2}$ represents the probability of a head-up turn."

"It's simple with a coin," someone interrupted. "Do it with something more complicated—a die, for instance."

"All right," the mathematician agreed. "Let's take a die. It's cubical in shape, with numbers on each of its faces (fig. 57). Now, what's the probability, say, of the number 6 turning up? How many possible occurrences are there? There are six faces and, therefore, any of the numbers from 1 to 6 can turn up. For us, the favourable occurrence will be when it is 6. The probability in this case will be $\frac{1}{6}$."

"Is it really possible to compute the probability of any event?" one of the girls asked. "Take this, for instance. I've a hunch that the first person to pass our window will be a man. What's the probability that my hunch is correct?"

"The probability is $\frac{1}{2}$, if we agree to regard even a year-old baby boy as a man. There's an equal number of men and women on our earth."

"And what's the probability that the first *two* persons will be men?" another asked.

"Here computation will be more complicated. Let's try all the possible combinations. First, it's possible that they will be men. Second, the first may be a man and the second a woman. Third, it may be the other way round: first the woman and then the man. And fourthly, both of them may be women. So, the number of possible combinations is 4. And of these combinations only one is favourable—the first. Thus, the probability is $\frac{1}{4}$. That's the solution of your problem."

"That's clear, but then we could have a problem of *three* men. What's the probability in this case that the first *three* to pass our window will be men."

"Well, we can calculate that too. Let's start with computing the number of possible combinations. For two passers-by the number of combinations, as we have seen, is 4. By adding a third passer-by we double the number of possible combinations because each of those 4 groups of two passers-by can be joined either by a man or a woman. Therefore, the number of possible combinations in this case will be $4\times2=8$. The obvious probability will be $\frac{1}{8}$, since only one combination will be the one we want. It's easy to remember the method of computing the probabilities: in the case of two passers-by the probability is $\frac{1}{2}\times\frac{1}{2}=\frac{1}{4}$; for three it is $\frac{1}{2}\times\frac{1}{2}\times\frac{1}{2}=\frac{1}{8}$; for four the probability will be equal to the product of 4 halves, etc. The probability as you may see, grows less each time."

"Then what will it be for 10 passers-by?"

"You mean what is the probability that the first ten passers-by are men? For that we have to find the product of 10 halves. That will be $\frac{1}{1024}$. That means if you bet a ruble that it will happen, I can wager 1,000 rubles that it will not."

"The bet is tempting!" one of those present exclaimed. "I'm more than willing to put up a ruble to win a thousand."

"But don t forget that the chance to win is one in a thousand."

"I don't care. I'd even bet a ruble against a thousand that the first hundred passers-by are all men."

"D'you realize how little the probability is in this case?"

"It's probably one in a million or something like that."

"No, it's immeasurably less. You'd have one in a million for 20 passers-by. For 100 we'd have. . . . Wait, let me calculate it on a sheet of paper. For 100 the probability would be—oh-ho—1/1,000,000,000,000,000,000,000,000,000,000!"

"Is that all?"

"You find that too little? Why, there aren't that many drops of water in an ocean, not even a 1,000 times less."

"Yes, the number *is* imposing! Well, how much will you put up against my ruble?"

"Ha, ha! Everything! Everything I have."

"Everything? That's too much. Make it your bicycle. Though I'm sure you won't dare."

"I won't dare? Go ahead. I bet you my bicycle. I'm not risking anything anyway."

"Neither am I. A ruble isn't much! And I stand to win a bicycle and you'll win, if you do, almost nothing."

"But don't you realize that you'll never win? You'll never get the bicycle and I've as good as got your ruble in my pocket."

"Don't do it," the mathematician's friend joined in. "It's madness to bet a bicycle against a ruble."

"On the contrary," the mathematician replied. "It's madness to bet even one ruble in such circumstances. It's a sure loss! That's plain throwing money away."

"But still there is a chance, isn't there?"

"Yes, a drop in an ocean. In ten oceans, in fact. That s how big the chance is. I'm betting ten oceans against that one chance. I'm just as sure to win as I'm sure that two and two is four."

"You're letting your imagination run away with you," an old professor broke in.

"What, professor, you really think he has a chance?"

"Have you considered the fact that not all occurrences are equally possible? When is computation of the probability of a coincidence correct? For equally possible events, isn't it? And here we have a. . . . But listen. I think you'll see your mistake now. D'you hear the military band?"

"I do. What has it got to do. . . ." The young mathematician stopped short. There was an expression of fear on his face as he rushed to the window.

"Yes," he said mournfully. "I've lost the bet. Bye-bye, bicycle."

A moment later we saw a battalion of soldiers marching past our window!

58. NUMBER GIANTS AROUND AND INSIDE US.—There is no need to go out of one's way to find number giants. They are all around us and even inside us—all one has to do is know how to recognize them. The sky above, the sand under our feet, the air around us, the blood in our body—all this conceals number giants.

For most people there is no mystery about number giants in space. Be it the number of stars in the sky, their distance from one another and the Earth or their size, weight or age—in each case we invariably come up against numbers that dwarf our imagination. It is not for nothing that people have coined the expression "astronomic number." But some people do not even suspect that some heavenly bodies that the astronomers call "little" are real giants when regarded from man's point of view. Our solar system has some planets only a few kilometres in diameter, and the astronomers, who are accustomed to deal with number giants, call them "tiny." But they are tiny only when compared to other heavenly bodies that are bigger: from our point of view they are far from being small. Let us take, for example, a recently discovered planet three kilometres in diameter. It is not difficult to calculate geometrically that its surface is equal to 28 square kilometres or 28,000,000 square metres. One square metre is enough space for seven persons standing upright. So, you see, there is enough space on the surface of this "tiny" planet for 196,000,000 persons.

The sand that we tread upon also introduces us to the world of number giants. It is not for nothing that there is the expression "as numerous as the grains of sand on the seashore." Incidentally, the ancients underestimated the number of grains of sand—they thought there were as many of them as stars in the sky. In the old days there were no telescopes and without them all man can see in one hemisphere is about 3,500 stars. The grains of sand on the seashore are millions of times more numerous than the stars one can see with the bare eye.

There is also a number concealed in the very air that we breathe. Each cubic centimetre, each thimble contains 27,000,000,000,000,000,000 molecules.

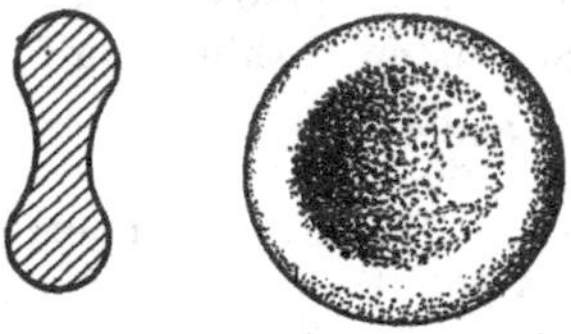

Fig. 58. A red corpuscle.

It is impossible even to imagine how big this number is. If there were as many people on Earth, there would not be enough space for them. Indeed, the surface area of the globe, counting all the continents and oceans, is equal to 500 million square kilometres. If we break this up into square metres, we get

500,000,000,000,000 square metres.

Now let us divide 27,000,000,000,000,000,000 by this number. The result is 54,000. And that means that there would be over 50,000 persons to every square metre!

We have said that every human being carries within himself a number giant—blood. If we examine a drop of blood under the microscope we shall see a huge number of red corpuscles. They look like disks flattened at the centre (fig. 58). They are all of approximately the same size—0.007 millimetre in diameter and 0,002 millimetre thick. There are a great many of them—5,000,000 in a tiny drop of blood of about 1 cubic millimetre. How many are there in a man's body? There are 14 times fewer litres of blood in a man's body than kilogrammes in his weight. For instance, if he weighs 40 kilogrammes, he has about 3 litres (or 3,000,000 cubic millimetres) of blood. A simple calculation will show that he has

$$5,000,000 \times 3,000,000 = 15,000,000,000,000 \text{ red corpuscles.}$$

Just think! 15,000,000 million red corpuscles! How long will a chain of these corpuscles be? That is not difficult to calculate: 105,000 kilometres, long enough to wind around the Earth's equator

$$100,000 : 40,000 = 2.5 \text{ times.}$$

And if we take a man of average weight, the chain of red corpuscles will be long enough to do that 3 times.

Fig. 59 Long enough to wind around the Earth's equator three times.

These tiny red corpuscles play an important role in our organism. They carry oxygen to all parts of the body. They absorb it when the blood passes through the lungs and then excrete it when the bloodstream drives them into the tissue of our body, into parts that are the farthest from the lungs. The smaller the corpuscles and the more numerous, the better they fulfil their function because then they have a greater surface and it is only through their surface that they can absorb and excrete oxygen. Calculation has shown that their total surface is many times greater than the surface of man's body; it is equal to 1,200 square metres—the size of a garden plot 40 metres long and 30 metres wide. Now you understand how important it is for the living organism to have as many as possible of these red corpuscles—they absorb and excrete oxygen on a surface that is 1,000 times bigger than our body.

Another number giant is the impressive total of the food consumed by a human being (taking 70 years as an average life span). It would take a regular freight train to transport all the tons of water, bread, meat, game, fish, vegetables, eggs, milk, etc., that one consumes in one's lifetime. It is difficult indeed to believe that a man can swallow—though not all at once, of course—a whole trainful of food.

CHAPTER VII

WITHOUT INSTRUMENTS OF MEASUREMENT

59. CALCULATING DISTANCE BY STEPS.—We do not always have a yard-stick with us and it is useful to know how to measure, even if only approximately.

The easiest way of measuring some distance, say, when you are out on a hike, is by steps. For that you must know the width of your step. Of course, your steps are not always of the same width. On the whole, however, they are more or less of the same width and if you know it, you can calculate any distance.

First you must calculate the average width of your steps. That, of course, cannot be done without an instrument of measurement.

Take a tape, stretch it out some 20 metres, take it away and then see how many steps you need to cover the distance. It is possible that the result will be x plus a fraction. If the fraction is less than half, do not count it at all; if it is more than half, count it as a whole. After that divide the 20 metres by the number of steps and get the average width. Memorize the result.

In order not to lose track of steps—especially when measuring a long distance—it is best to count up to 10 and then bend in one finger of your left hand. When all the fingers are bent, i.e., when you have covered 50 steps, you bend in one finger of your right hand. Thus, you can count up to 250, and then start all over again. Only you must not forget how many times in all you have bent the fingers of your right hand. If, for instance, you have reached your destination and have twice bent in all the fingers of your right hand and have another three fingers bent on the right and four on the left, it means you have made

$$2\times250+(3\times50)+(4\times10)=690 \text{ steps.}$$

To this total you must, of course, add the few steps you have made after bending in the last finger of your left hand, if such is the case.

By the way, here is an old rule: the average width of an adult's step is equal to half the distance from his eye to his toe.

Another old rule applies to the *speed* of walking: a man does as many kilometres in an hour as he does steps in three seconds. But this rule is correct only for a certain width of step, and a big step at that. In fact, if the width of the step is x metres and the number of steps in three seconds is n, then in three seconds a man covers nx metres and in an hour (3,600 seconds) 1,200 nx metres or 1.2 nx kilometres. If this distance is to equal the number of steps made in 3 seconds, there must be the following equation:

Fig. 60. The distance between the tip of an outstretched arm to the opposite shoulder is about one metre.

$$1.2\ nx = n$$

or

$$1.2\ x = 1,$$

hence

$$x = 0.83 \text{ metres}$$

The rule that the width of a man's step depends on his height is correct; the second rule—the one we have just examined—applies only to men of average height, i.e., men who are about 1.75 metres tall.

60. A LIVE SCALE.—When there is no instrument of measurement around, the following is a good way of measuring average-sized objects. Stretch a string or a stick from the tip of an outstretched arm to the opposite shoulder (fig. 60). In the case of an adult this distance is about one metre long. Another way of measuring a metre (approximately) is with one's fingers: the distance between the index finger and the thumb stretched as wide apart as possible is about 18 centimetres and six of such distances make approximately 1 metre (fig 61a).

This teaches us to measure with "bare hands." The only thing one need know for that is the size of one's palm, and remember it.

First one must know the width of one's palm, as shown in fig 61b. In the case of an adult it is usually 10 centimetres. Yours may be smaller or bigger; you must know by how much. Then you must know the distance between the index and middle fingers, stretched as wide apart as possible (fig 61c). It is also useful to know the length of the index finger, from the base of the thumb (fig 61d). And, finally, calculate the space between the thumb and the little finger when extended (fig. 61e).

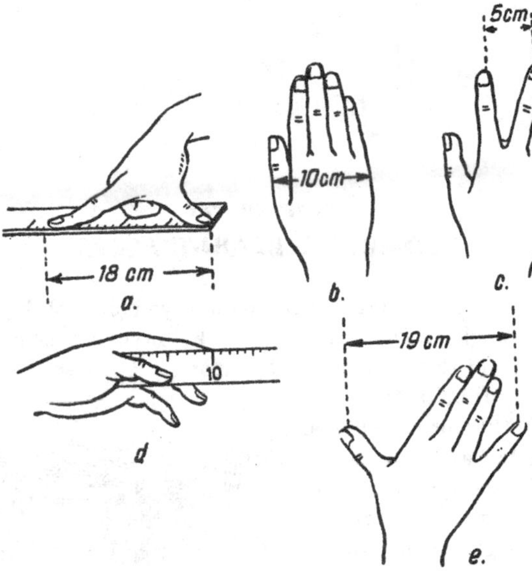

Fig. 61. Measuring with the hand.

Making use of these "live scales," you can obtain approximate measurements of small-sized objects.

CHAPTER VIII

GEOMETRIC BRAIN-TEASERS

To solve the conundrums in this chapter you do not have to know geometry thoroughly. That can be done by anyone possessing elementary knowledge of this branch of mathematics. The two dozen problems offered here will help the reader to check whether or not he really knows geometry as he thinks he does. Real knowledge does not mean just knowing how to describe the peculiarities of geometric forms, but how to apply them to the solution of practical problems. Of what use is a gun to a man if he does not know how to shoot?

Fig. 62. Why does the front axle wear out

Let the reader see for himself how many bull's-eyes he can score out of 24 shots at these geometric targets.

61. A CART.—Why does the front axle of a cart wear out faster than the rear?

62. THROUGH A MAGNIFYING GLASS.—How big will the angle of $1\frac{1}{2}°$ seem if you look at it through a glass that magnifies things four times (fig. 63)?

Fig. 63. How big will the angle seem?

63. A CARPENTER'S LEVEL.—You have probably seen a carpenter's level with a glass tube with a bubble (fig. 64) that deviates from the centre when placed on a sloping surface. The bigger the slope, the more does the bubble deviate from the mark. It moves because, being lighter than the liquid in the tube, it rises to the surface. If the tube were straight, the bubble would move to the end of the tube, that is, to its highest point. A level like that, as it may easily be seen, would be very inconvenient. That is why the tube is usually arched, as shown in fig. 64. When the level is horizontal, the bubble, situated at the highest point of the tube, is in the centre; if the level is sloped, the highest point is then not its centre, but some point next to it, and the bubble moves from the mark to another part of the tube.*.

Fig. 64. The carpenter's level.

The problem is to determine how many millimetres the bubble will move away from the mark if the level is sloped $\frac{1}{2}^\circ$ and the radius of the arch of the tube is 1 metre.

64. HOW MANY EDGES?—Here is a question that will probably sound either too naive or, on the contrary, too tricky.

How many edges has a hexagonal pencil?

Think well before you look at the answer.

65. A CRESCENT.—Can you divide a crescent (fig. 65) into six parts by drawing just two straight lines?

66. A MATCH TRICK.—Out of 12 matches you can build the figure of a cross (fig. 66), the area to equal five "match" squares.

Can you rearrange the matches in such a way as to cover an area equal to only four "match" squares?

The use of measuring instruments is forbidden.

Fig. 65. A crescent.

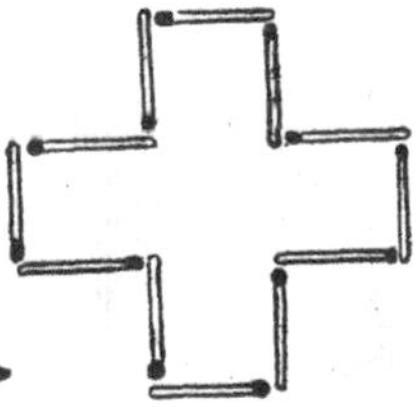

Fig. 66. A cross of twelve matches.

67. ANOTHER MATCH TRICK.—Out of eight matches you can make all sorts of figures. Some of them are shown in fig. 67. They are all

* It would be more correct to say that "the mark moves from the bubble," because the latter really remains in its place while the tube and the mark glide past.

different in size. The task is to make the *biggest* possible figure out of these eight matches.

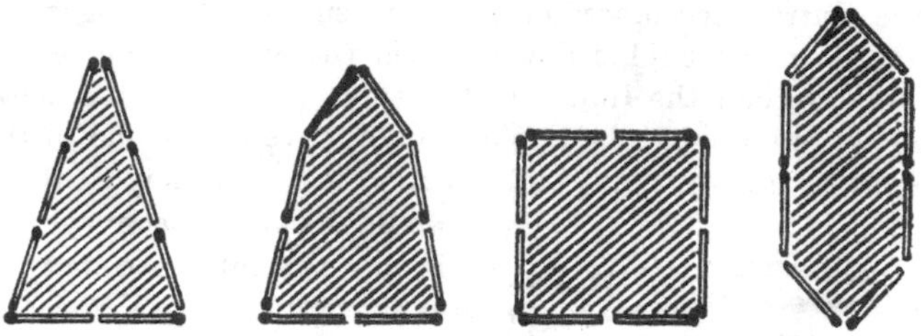

Fig. 67. How to make the biggest possible figure out of eight matches.

68. THE WAY OF THE FLY.—On the wall inside a cylindrical glass container, three centimetres from the upper circular base, there is a drop of honey. On the lateral surface, diametrically opposite it, there is a fly (fig. 68).

Show the fly the shortest route to the honey.

The diameter of the cylinder is 10 centimetres and the height 20.

Don't expect the fly to find this way itself and thus facilitate the solution of the problem: for that it would have to be well versed in geometry, and that is something beyond a fly's ability.

Fig. 68. Show the fly the short cut to honey.

69. FIND A PLUG.—You are given a small plank (fig. 69) with three holes: square, triangular and circular. Can you make one plug that would fit all the three apertures?

70. THE SECOND PLUG.—If you have solved the previous problem, try to find a plug that would close the apertures shown in fig. 70.

71. THE THIRD PLUG.—And here is yet another problem of the same type. Find a plug for the three apertures in fig. 71.

72. A COIN TRICK.—Take a couple of coins—5-kopek and 2-kopek (any two similar coins of 18 mm. and 25 mm. in diameter will do). Then, on a sheet of paper, cut out a circle equal to the circumference of the 2-kopek coin.

Do you think the 5-kopek coin will get through this hole?

There is no catch to the problem; it is genuinely geometrical.

Fig. 69. Find a plug for these three apertures.

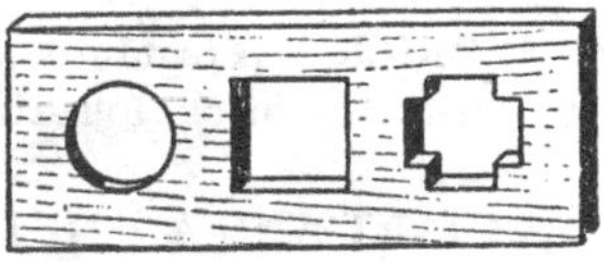

Fig. 70. Is there one plug for all three apertures?

73. THE HEIGHT OF A TOWER.—There is a very big tower in your town, but you do not know its height. You have, however, a photograph of the tower. Can it help you to find the real height?

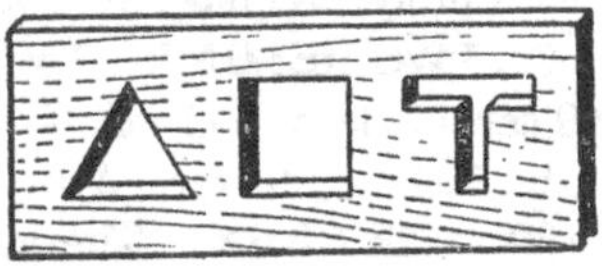

Fig. 71. Can you make one plug fit all three of these apertures?

74. SIMILAR FIGURES.—This problem is meant for those who understand geometrical similarity. Answer the following two questions:

1) Are the two triangles in fig. 72 similar?

2) Are the outer and inner rectangles of the picture frame in fig. 73 similar?

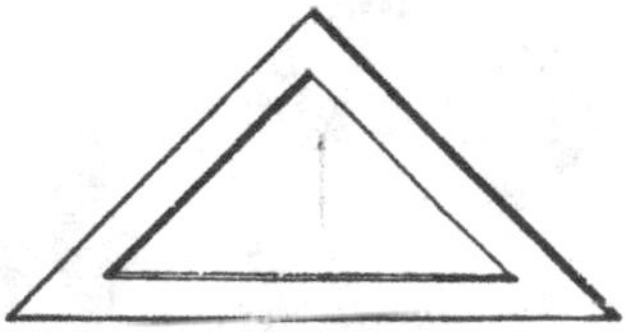

Fig. 72. Are those two triangles similar?

75. THE SHADOW OF A WIRE.—How far, on a sunny day, does the perfect shadow of a wire 4 mm. in diameter stretch?

Fig. 73. Are the inner and outer rectangles similar?

76. A BRICK.—A regular-size brick weighs 4 kilogrammes. How much will a similar toy brick, four times smaller but made of the same material, weigh?

77. A GIANT AND A PYGMY.—By how much does a man 2 metres tall outweigh a pygmy who is only 1 metre tall?

78. TWO WATER-MELONS.—A man is selling two water-melons. One is one-quarter bigger than the other, but costs one and a half times more. Which one would you buy (see fig. 74)?

79. TWO MELONS.—Two melons of the same sort are being sold. One is 60 centimetres in circumference, and the other is 50. The first is one and a half times dearer. Which of the two is it more profitable to buy?

80. A CHERRY.—The pulp of a cherry around the stone is as thick as the stone itself. Let us assume that the cherry and the stone are round. Can you calculate mentally how much more pulp than stone there is in the cherry?

Fig. 74. Which water-melon is it better to buy?

81. THE EIFFEL TOWER.—The 300-metre-high Eiffel Tower in Paris is made of steel—8,000,000 kilogrammes of it. I have decided to order a model of this tower, one weighing a kilogramme.

How high will it be? Will it be bigger or smaller than a drinking glass?

82. TWO PANS.—There are two pans that are similar in form and of the same thickness; one of them is eight times more capacious than the other.

How much heavier is it than the smaller one?

83. IN WINTER.—An adult and a child, similarly dressed, are standing in the street on a wintry day.

Who feels colder?

84. SUGAR.—What is heavier: a glass of powdered sugar or a similar glass of lump sugar?

Answers 61 to 84

61. At the first glance this problem does not look geometrical at all. But one who knows geometry well will know how to find a geometrical basis where it is disguised by all sorts of extraneous details.

This problem is a geometrical one and without geometry it is impossible to solve it.

And so, why does the front axle wear out faster than the rear? If you look properly at fig. 62 you will see that the front wheels are smaller than the rear. Geometry teaches us that a circle with a smaller circumference has to make more revolutions than a bigger circle to cover the same distance. And it is only natural that the more the wheel turns, the quicker the front axle wears out.

62. If you think that the magnifying glass increases our angle to $1\ 1/2 \times 4 = 6°$, you are very much mistaken. The magnifying glass does not increase the magnitude of the angle. True, the arc measuring the angle increases, but then its radius increases proportionally too, and the result is that the magnitude of the central angle remains unchanged. Fig. 75 explains this.

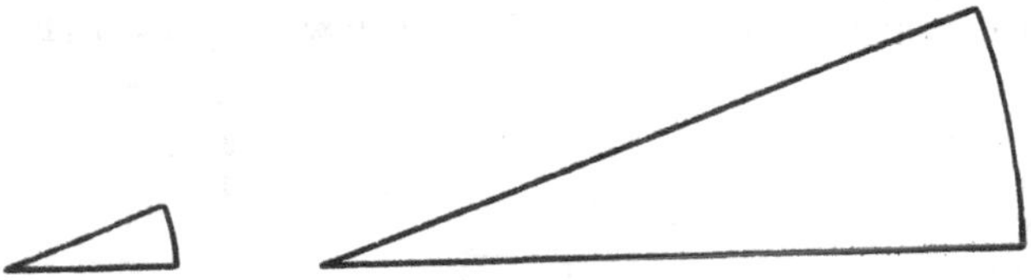

Fig. 75.

63. In fig. 76 *MAN* is the original position of the level's arc, *M'BN'* in the new position with the chord *M'N'* and the chord *MN* forming an angle of 1/2°. The bubble, formerly at *A*, remains at the same point but the centre of the arc *MN* has moved to *B*. We must now calculate the length of the arc *AB* with the radius being equal to 1 metre and the magnitude of the angle to 1/2° (this follows from the fact that we are dealing with corresponding acute angles with perpendicular sides).

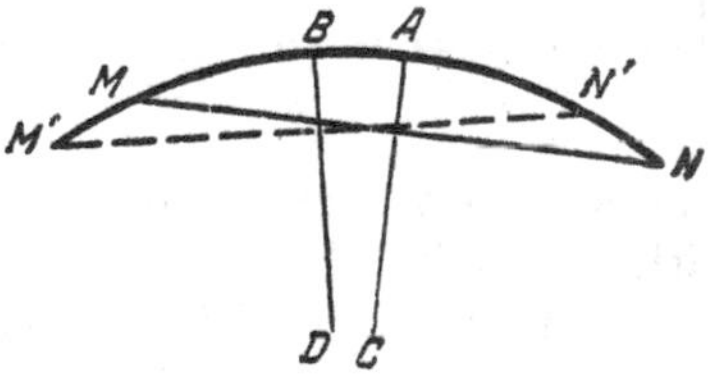

Fig. 76.

It is not difficult to calculate that. Since the radius is equal to 1 metre (1,000 millimetres), the circumference will equal $2 \times 3.14 \times 1{,}000 =$ 6,280 millimetres. And since there are 360° or 720 half-degrees in a circumference, the length of 1/2° in this particular case will be:

$$6{,}280 : 720 = 8.7 \text{ millimetres}$$

Thus, the bubble will move from the mark (or rather the mark will move from the bubble) by approximately 9 millimetres. It is obvious

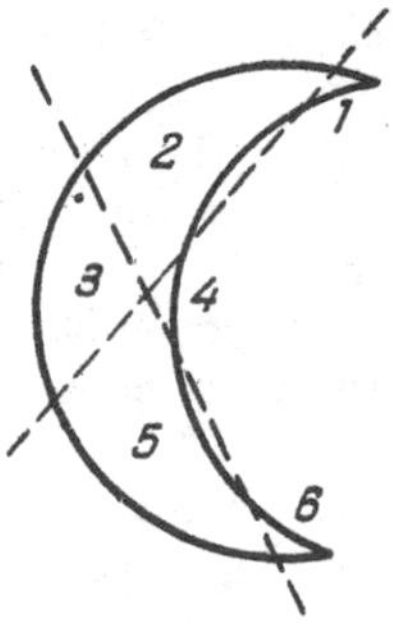

Fig. 77.

that the greater the radius of the curvature of the tube, the more sensitive is the level.

64. There is nothing tricky in this problem. The only catch is the erroneous interpretation of meaning. A "hexagonal" pencil has not six edges as most people probably think. If not sharpened, it has eight: six faces and two small bases. If it really had only six edges, it would have a different form altogether—the form of a rectangle.

65. This should be done as shown in fig. 77. The result is six parts which are numbered for convenience's sake.

66. The matches should be laid out as shown in fig. 78*a*. The area of the figure is equal to the quadrupled area of a "match" square. It is quite obvious that this is so. Let us mentally fill out our figure to form a triangle. What we get is a right triangle

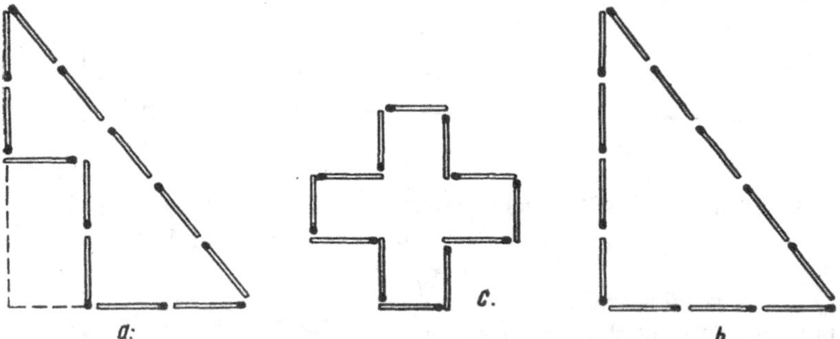

Fig. 78.

whose base is equal to three matches and its altitude to four.* Its area is equal to one-half of its base times its altitude: $^1/_2 \times 3 \times 4 = 6$ (fig. 78*b*) "match" squares. But the area of our figure is obviously smaller than the area of the triangle by two "match" squares and is therefore equal to four such squares.

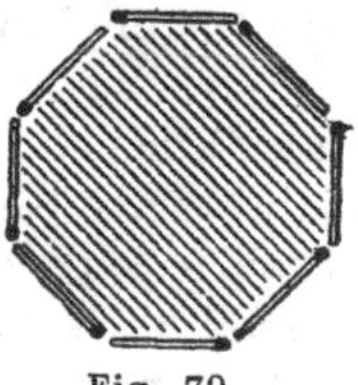
Fig. 79.

67. It can be proved that of all the closed plane figures the circle is the biggest. It is, of course, impossible to make one out of matches. However, out of eight matches it is possible to make a figure (79) that most closely resembles a circle—a regular octagon.

* Readers who are acquainted with the Pythagorean Proposition will understand why we are so certain that ours is a right triangle: $3^2+4^2=5^2$.

And this regular octagon is precisely the figure that we require, for it is the biggest in area.

68. To solve this problem we must slit the cylindrical container open and flatten out the surface. The result will be a rectangle (fig. 80) whose width is 20 centimetres and whose length is equal to the circumference, i. e., $10\times 3\frac{1}{7}=31.5$ centimetres (approx.). Now let us mark in this rectangle the position of the fly and that of the drop of honey. The fly is at point A, 17 centimetres from the base, while the drop of honey is at point B, at the same height but half the circumference of the cylinder away from A, that is, $15\frac{3}{4}$ centimetres away.

To find the point where the fly must climb over into thec ylinder we must do as follows. From point B (fig. 81) we shall draw a perpendicular line to the upper base and continue it up to a similar distance. Thus, we shall obtain C, which we shall connect by a straight line

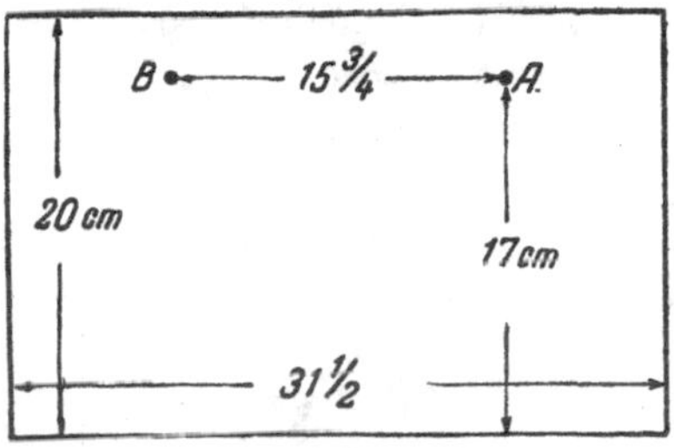

Fig. 80.

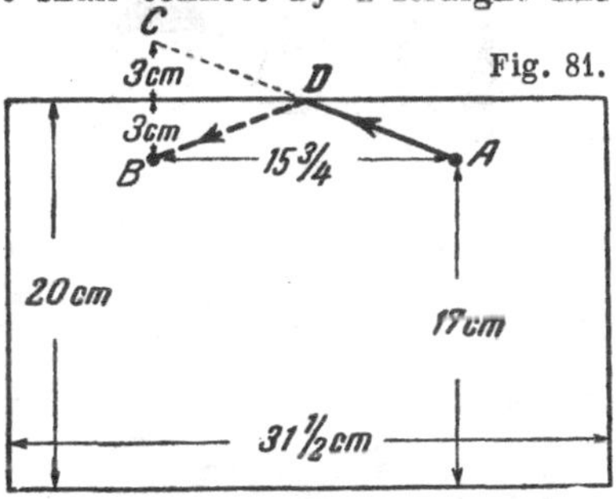

Fig. 81.

with A. Point D will be the one where the fly should cross over into the cylinder, and the route ADB is the shortest way.

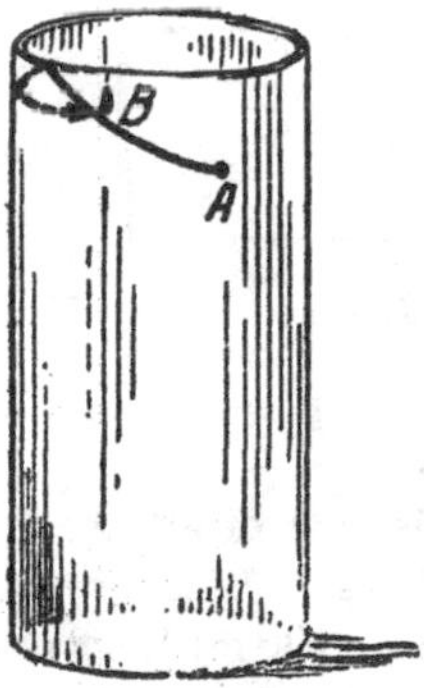

Fig. 82.

Having found the shortest route on a flattened rectangle we can roll it back into a cylinder and see how the fly must travel to reach the drop of honey (fig. 82).

I don't know whether or not this is the route taken by flies in such cases. It is possible that, possessing a good nose, flies actually use this shortest route—possible but not probable. A good nose is not enough without knowledge of geometry.

69. There is such a plug. It is shown in fig. 83 and, as you may see, it can really close all the three apertures: square, triangular and circular.

70. There is also a plug to close the three apertures in fig. 84: circular, square and cruciform. It is shown in all its three aspects.

71. Finally, there is a plug like that too. You may see all its aspects in fig. 85.

72. Strange as it may seem, it is quite possible to pass a 5-kopek coin through such a small hole. The paper is folded so that the circle is stretched out into a slit (fig. 86), and it is through this slit that the 5-kopek coin passes.

Geometry easily explains this seemingly tricky phenomenon. The 2-kopek coin is 18 millimetres in diameter. It is not difficult to calculate its circumference: it is slightly over 56 millimetres. The length of the slit, therefore, is half of that, or 28 millimetres. And since a 5-kopek coin is 25 millimetres in diameter, it can easily pass through a 28-millimetre slit, even if it is 1.5 millimetres thick.

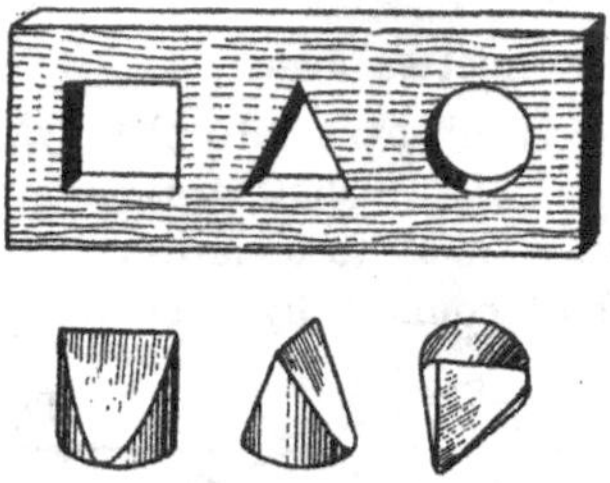

Fig. 83.

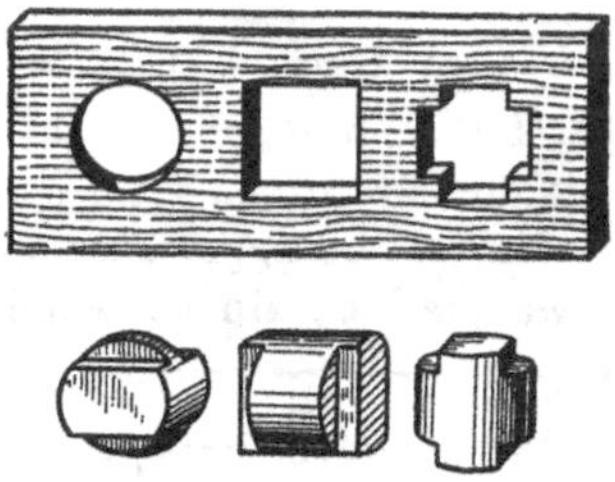

Fig. 84.

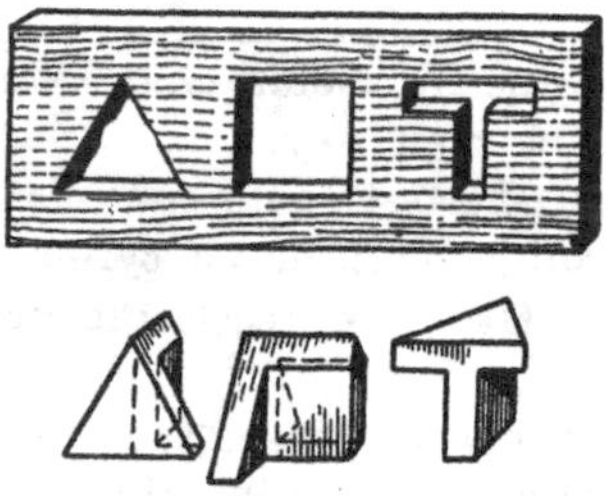

Fig. 85.

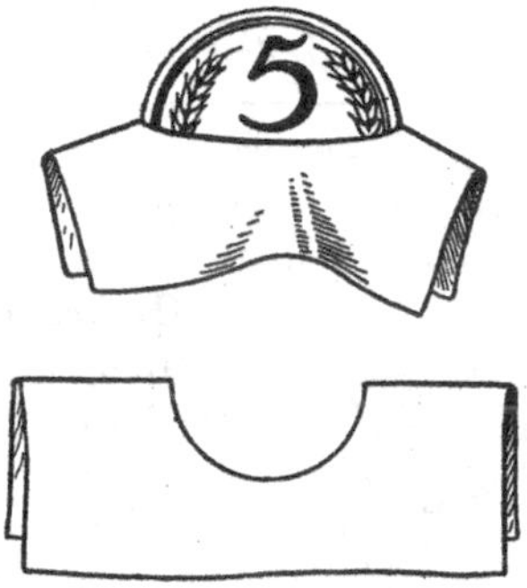

Fig. 86.

73. To determine the real height of the tower, it is first of all necessary to have the correct measurements of its height and base in the photograph. Let us assume that they are 95 and 19 millimetres respectively. After that you measure the base of the real tower. Let us suppose that it is 14 metres wide.

Geometrically, the tower in the photograph and the real tower are proportionally the same, i.e., the ratio between the height and the base of the tower in the photograph is equal to that between the height and the base of the real tower. In the first case it is 95 : 19, i.e., 5. Hence, the height of the tower is 5 times greater than the base. Therefore, the height of the real tower is:

$$14 \times 5 = 70 \text{ metres}$$

There is a "but," however To determine the height of a tower you must have a really good photograph, not a distorted one—the kind inexperienced amateur photographers sometimes take.

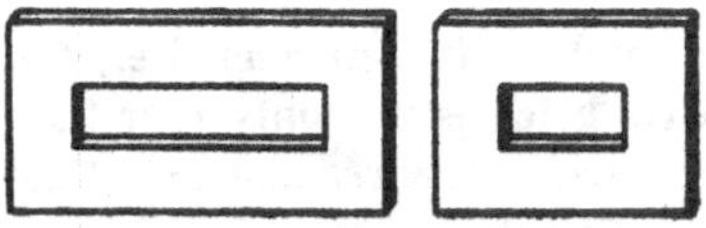

Fig. 87.

74. Very often both of these questions are answered in the affirmative. In reality it is only triangles that are similar. The outer and inner rectangles of the picture frame, generally speaking, are not similar. For triangles to be similar it is enough for their angles to be correspondingly equal; and since the sides of the inner triangle are parallel to the sides of the outer, the figures are similar. As for the similarity of the polygons, it is not enough that their angles be equal (or and that is the same thing—that their sides be parellel): it is also necessary for the *sides* of the polygons to be *proportional*. As far as the outer and inner rectangles of a picture frame are concerned such is the case only with squares (and rhombi generally). In all other cases the sides of the outer and inner rectangles are not proportional and the figures, therefore, are not similar. The absence of similarity becomes all the more obvious in thick rectangular frames (fig. 87). In the frame on the left the outer sides are in the ratio of 2 : 1 and the inner 4 : 1. In the frame on the right, they are 4 : 3 and 2 : 1 respectively.

75. Many people will be surprised to learn that the solution of this problem requires knowledge of astronomy: of the distance between the Earth and the Sun and of the size of the Sun's diameter.

The length of the perfect shadow cast by a wire is determined by the geometric figure shown in fig. 88. It is easy to see that the shadow is as many times greater than the diameter of the wire as the distance between the Earth and the Sun (150,000,000 kilometres) is greater than the Sun's diameter (1,400,000 kilometres). In round figures, the ratio in the latter case is 115. Therefore, the perfect shadow cast by the wire stretches:

$$4 \times 115 = 460 \text{ millimetres} = 46 \text{ contimetres.}$$

The insignificant length of a perfect shadow explains why it is not always seen on the ground or house walls; the weak streaks that one does see are not shadows, but penumbra.

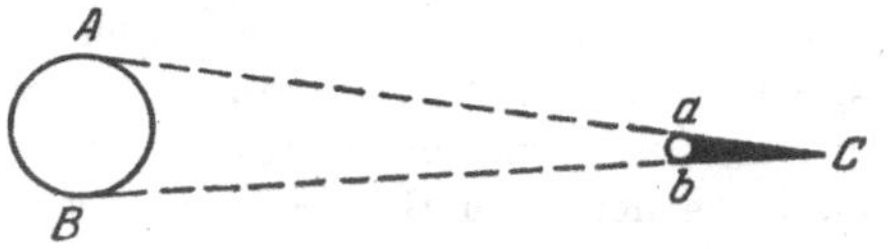

Fig. 88.

Another method of solving such problems was shown in brain-teaser 7.

76. The answer that the toy brick weighs 1 kilogramme, i.e., four times less, is absolutely wrong. The little brick is not only four times *shorter* than the real one, but also four times *narrower* and four times *lower*, and its volume and weight are therefore $4\times4\times4=64$ times less. The correct answer therefore would be:

$$4{,}000 : 64=62.5 \text{ grammes.}$$

77. This problem is similar to the one above, so you should be able to solve it correctly. Since human bodies are more or less similar, the man who is twice taller outweighs the other not two, but eight times.

The biggest man the world knows of was an Alsatian 2.75 metres tall—approximately 1 metre taller than a man of average height. And the smallest was a lilliputian less than 40 centimetres tall, i.e., roundly speaking, he was seven times shorter than the Alsatian. If we were to weigh the two, we would have to put on the other pan of the balance: $7\times7\times7=343$ lilliputians, and that's a whole crowd.

78. The size of the big water-melon exceeds that of the small one

$1\frac{1}{4}\times1\frac{1}{4}\times1\frac{1}{4}=\frac{125}{64}$, or almost twice.

Therefore, it is better to buy the big one: it costs only one and a half times more and has over two times more pulp.

Why then, you may ask, should the vendors demand only one and a half times more for such water-melons and not twice? The explanation is simple: most vendors are weak in geometry. But for that matter so are the buyers, and that is the reason why they often refuse such profitable deals. It can be definitely affirmed that it is better to buy big water-melons than small ones because they are always priced less than what they really should cost—but most buyers do not even suspect that.

And for the same reason it is more profitable to buy big eggs than small ones, that is, if they are not sold by weight.

79. Circumferences are to one another as their diameters. If the circumference of one melon is 60 centimetres and of the other 50 centimetres, then the ratio between their diameters is $60 : 50=\frac{6}{5}$, and the ratio between their sizes is:

$$\left(\frac{6}{5}\right)^3=\frac{216}{125}=1.73$$

The bigger melon, if it were priced according to its size (or weight), should cost 1.73 times or 73 per cent more than the small one. Yet the vendor asks only 50 per cent more. It is obvious, therefore, that it is more profitable to buy the bigger one.

80. The conditions of the problem say that the diameter of the cherry is three times that of the stone. Hence, the size of the cherry is $3\times3\times3=27$ times that of the stone. That means that the stone occupies $\frac{1}{27}$ part of the cherry and the pulp the remaining $\frac{26}{27}$. In other words, the pulp is 26 times bigger in volume than the stone.

81. If the model is 8,000,000 times lighter than the real Eiffel Tower and both are made of the same metal, then the *volume* of the model should be 8,000,000 times less than that of the real tower. We already know that the volumes of similar figures are to one another as the cubes of their altitudes. Hence, the model must be 200 times smaller than the original because

$$200\times200\times200=8,000,000$$

The altitude of the real tower is 300 metres. Therefore, the height of the model should be

$$300 : 200=1\frac{1}{2} \text{ metres.}$$

The model will thus be about the height of a man.

82. Both pans are geometrically similar bodies. If the bigger one is eight times more capacious, then all its linear measurements are two times greater: it is twice bigger in height and breadth. But if it is the case, then its surface is $2\times2=4$ times greater because the surfaces of similar bodies are to one another as the squares of their linear measurements. Since the wallsides are of the same thickness, the weight of the pan depends on the size of its surface. Hence, the answer: the bigger pan is *four times* heavier.

83. At the first glance this problem does not look mathematical at all, but in fact, like the previous one, it is solved geometrically.

Before we set out to solve this problem let us examine another one—of the same kind but simpler.

Two boilers, one bigger than the other, made of the same material and similar in form, are filled with hot water. In which of the two will the water cool down faster?

Things usually cool down from the surface. Therefore, the boiler with a bigger surface per unit of volume cools down faster. If one of the boilers is n times higher and broader than the other, then its surface is n^2 times greater and the volume n^3 times bigger; for each unit of the surface in the big boiler there are n times more volume. Hence, the smaller boiler cools down faster.

For the same reason a child standing out in the street on a wintry day feels the cold more than a similarly dressed adult: the amount of heat in each cubic centimetre of the body is approximately the same in the case of both, but a child has a greater cooling surface per one cubic centimetre of the body than an adwlt.

That is the reason why man's fingers and nose suffer more from cold and get frost-bitten oftener than any other parts of the body whose surface is not so great when compared to their volume.

And, finally, that also explains, for instance, the following problem: Why does splint wood catch fire faster than the log from which it has been chopped off?

Since heat spreads from the surface to the whole volume of a body, it is necessary to compare the surface and volume of splint wood (for instance, square section) with the surface and volume of a log of the same length and same square section in order to determine the size of the surface per one cubic centimetre of wood in both cases. It the log is ten times thicker than splint wood, then the lateral surface of the log is ten times bigger than that of splint wood and its volume 100 times. Therefore, for each unit of the surface of splint wood there is ten times less volume than in a log: the same amount of heat heats ten times less material in splint wood. Hence, the same source of heat sets splint wood on fire faster than a log. (Because of the poor heat conductibility of wood the comparison should be regarded as only roughly approximate—it is characteristic of the whole process and not of the quantitative aspect.)

84. A little bit of brainwork will show you that this seemingly tricky problem is actually quite simple. Let us assume for simplicity's sake that the diameter of lump sugar is 100 times that of powdered sugar. Let us then imagine that the diameter of the sugar particles and the glass which they fill increase 100 times. The capacity of the glass will increase $100 \times 100 \times 100 = 1{,}000{,}000$ times. The weight of the sugar in it will increase proportionally. Let us next measure out an ordinary glass of such enlarged powdered sugar, i.e., one-millionth part of the contents of our giant glass. It will naturally weigh exactly the same as an ordinary glass of ordinary powdered sugar. What then does our enlarged powdered sugar represent? Simply a lump of sugar. Therefore, a glass of lump sugar weights the same as a glass of powdered sugar.

It would not make any difference if we enlarged a sugar particle 60 times instead of 100. The argument boils down to the fact that lump sugar is geometrically similar to powdered sugar. That assumption may not be 100 per cent correct, but it is very close to reality.

CHAPTER IX

THE GEOMETRY OF RAIN AND SNOW

85. PLUVIOMETER.—In the Soviet Union it has become a rule to consider Leningrad a very rainy city, far more rainy, for instance, than Moscow. But scientists deny that. They claim that rain brings more water to Moscow than to Leningrad. How do they know that? Is there really a way of measuring rain-water?

The task looks difficult, yet you can learn to do it yourself. Don't think that you have to collect all the water that descends to the ground. If rain-water did not spread and if it were not absorbed by soil, it would be enough to measure the *depth*. And that would not be difficult at all. When rain falls, it falls evenly everywhere: there is no such thing as watering one garden bed more than its neighbour. It is enough, therefore, to measure the depth at one spot to know the depth in the entire afflicted area.

Now you have probably guessed what you must do to measure the amount of rain-water. All you have to do is take a small lot where water would not spread or disappear underground. Any open vessel is suitable for that purpose, say, a bucket. If you have one with perpendicular wallsides (in the shape of a right circular cylinder), put it out when it rains.* When the rain stops, measure the depth of the water in the receptacle, and you will have everything you require for your computation.

Let us see how our home-made pluviometer works. How to measure the depth of the water in a bucket? With a ruler? That is not a bad way if there is plenty of water in it. But as a rule there are only 2 or 3 centimetres, and sometimes even a few millimetres, of water in the bucket, and in that case it is impossible to do it accurately. In our case every millimetre, in fact every fraction of it, is important. Well, what must we do?

* You should place your vessel as high as possible so that the drops falling on the ground do not ricochet into it.

The best thing is to pour the water from the bucket into some narrower glass vessel. Here the water will be at a higher level and it will be easy to see how high through the transparent wallsides. Of course, the depth of the water in the narrow receptacle will not be the depth we seek, but then it is easy to convert one measurement into another. If the diameter of the base of the narrow receptacle is ten times smaller than that of our bucket pluviometer, then the area of its base will be $10 \times 10 = 100$ times smaller than the area of the base of the bucket. It is clear that the water level in the glass receptacle will be 100 times higher than that in the bucket. Therefore, if there are 2 millimetres of rain-water in the bucket, there will be 200 millimetres, or 20 centimetres, in the glass receptacle.

From this calculation you will see that this receptacle should not be too much narrower than the bucket pluviometer, for then we would need an extremely high one to measure the depth of rain-water. Five times narrower will be good enough: then the area of its base will be 25 times smaller than the area of the bucket's base, and the level of the water will be 25 times higher. Each millimetre of water in the bucket will be equal to 25 millimetres in the glass receptacle. For convenience's sake paste a strip of paper on the outer side of the glass vessel and divide it into 27-millimetre sections, marking each of them 1, 2, 3, etc. Looking at the height of the water in the glass receptacle, you will know right away its depth in the bucket pluviometer without having to do any converting. If the diameter of the glass receptacle is four, and not five, times smaller than that of the bucket, then the sections on the paper strip should be 16 millimetres wide.

It is extremely inconvenient to pour water from a bucket into a narrow vessel. A good way out is to drill a hole in the wallside of the bucket and to draw the water out through a glass tube (on the pattern of a tap).

And so you have the necessary equipment to measure the depth of rain-water. A bucket and a home-made rain-gauge, naturally, are not as exact as a real pluviometer or the graduated glass used at meteorological stations. Still, this simple and cheap equipment will enable you to do many instructive calculations.

Here are a few problems.

86. HOW MUCH RAINFALL?—You have a kitchen garden 40 metres long and 24 metres wide. It has just stopped raining and you wish to know how much water has fallen on it. How is one to compute that?

You must start with determining the depth of rain-water: without knowing it you cannot do anything. Let us assume that your home-made pluviometer shows that there are 4 millimetres of rain-water. Let us next calculate how many cubic centimetres of water there will be on each square metre of the kitchen garden, that is, if the water were not absorbed by soil. One square metre means 100 centimetres in width and 100 centimetres in length. It is covered with 4 millimetres, i.e., 0 4 centimetres, of water. Hence, the volume of such a stratum of water would be:

$$100 \times 100 \times 0.4 = 4{,}000 \text{ cu. cm.}$$

You know that 1 cu. cm. of water weighs 1 gr. Therefore, on each square metre of the kitchen garden there are 4,000 gr. or 4 kg. of water. The area of your kitchen garden is:

$$40 \times 24 = 960 \text{ sq. m.}$$

That means the weight of rain-water on your kitchen garden is equal to:

$$4 \times 960 = 3{,}840 \text{ kg.}$$

or a little less than 4 tons.

Just for fun's sake, calculate how many buckets you would have to tote to give your kitchen garden the water brought it by rain. An ordinary bucket holds approximately 12 kg. of water. Therefore, the rain poured 3,840 : 12=320 buckets of water on your kitchen garden.

And so, you would have to empty more than 300 buckets to give your kitchen garden the amount of water rained down in some 15 minutes.

Can we express a shower or a drizzle in figures? For that it is necessary to determine how many millimetres of rain falls *in a minute*. If the rain is such that *every minute* there fall 2 millimetres of water, then it is a shower of extraordinary force. If it is an autumn drizzle, then it usually takes an hour and sometimes even longer for 1 millimetre of water to accumulate.

As you see, to measure the depth of rain-water is not only possible, but simple. More, you can, if you want to, determine even the number of raindrops, though that approximately.* In fact, in an ordinary rain there are on the average 12 drops to a gramme. Therefore, if we take the rain of which we spoke above, we shall see that there were:

* Rain always falls in drops, even when we think it is pouring.

$$4{,}000 \times 12 = 48{,}000 \text{ drops to a square metre.}$$

It is not difficult to calculate how many drops fell on the entire kitchen garden. But such computation, though interesting, is useless. The only reason we mentioned it was to show that it is possible to do the most incredible calculations, if you only know how to go about them.

87. **HOW MUCH SNOW?**—We have learned how to measure the depth of rain-water. What must we do to calculate the depth of water when it hails? Absolutely the same thing. Hail-stones fall into your rain-gauge and melt. After that you measure the depth.

It is a different case when it comes to snow-water. Here the pluviometer will not give you correct data because the wind blows part of the snow out of the bucket. But then it is possible to determine the depth of snow-water without a pluviometer. You can measure the depth of snow in a yard or in a field with the help of a wooden rod. And to find how deep the *water* will be when the snow melts, you have to do an experiment: fill a bucket with snow of equal friability, melt it and measure the depth of water. You will thus determine how many millimetres of water you get out of a centimetre of snow. Knowing that, you will not find it difficult to convert the depth of snow into the depth of water.

If, every day without fail, you measure the depth of rain-water during warm weather and add the water you obtain from snow in winter, you will know annual precipitation of your district.

Below are the average precipitation figures for a number of Soviet towns:

Leningrad	47 cm.
Vologda	45 "
Arkhangelsk	41 "
Moscow	55 "
Kostroma	49 "
Kazan	44 "
Kuibyshev	39 "
Chkalov	43 "
Odessa	40 "
Astrakhan	14 "
Kutaisi	179 "
Baku	24 "
Sverdlovsk	36 "
Tobolsk	43 "
Semipalatinsk	21 "
Alma-Ata	51 "
Tashkent	31 "
Yeniseisk	39 "
Irkutsk	44 "

Of these towns Kutaisi (179 cm.) gets the most water from the sky and Astrakhan (14 cm.) the least, 13 times less than Kutaisi. But there are places in the world where there is much more precipitation than in Kutaisi. For instance, there is a district in India which is virtually inundated by rain-water—the annual rainfall there is 1,260 cm., i. e., more than 12.5 metres! There was a day when rainfall there exceeded 100 cm. Then there are places where there is very much less rainfall than in Astrakhan—for instance, in Chile the figure is below 1 cm. *a year.*

Areas where rainfall is below 25 cm. are droughty, and without artificial irrigation agriculture is impossible.

It is easy to see that, having measured annual rainfall in various parts of the globe, it is possible to determine the annual average for the whole world. The annual average rainfall on land is 78 cm. It is said that there is approximately as much rainfall in the oceans as on land. Knowing that, it is not difficult to calculate the amount of precipitation—rain, hail, snow, etc.—upon the entire surface of the earth. For that you must know the latter's area. If you don't know it, here is how to compute it.

A metre is almost exactly 1/40,000,000th part of the globe's circumference. In other words, the circumference is 40,000,000 metres or 40,000 kilometres. The diameter of the globe is approximately $3\frac{1}{7}$ times smaller than its circumference. Knowing that, we can easily compute its diameter:

$$40{,}000 : 3\frac{1}{7} = 12{,}700 \text{ km.}$$

The rule governing the computation of the area of the surface of a sphere is as follows: multiply the diameter by itself and then by $3\frac{1}{7}$:

$$12{,}700 \times 12{,}700 \times 3\frac{1}{7} = 509{,}000{,}000 \text{ sq. km.}$$

(Starting with the fourth digit of the result, we write down noughts, since it is only the first three digits that are reliable.)

And so, the area of the surface of the globe is equal to 509 million sq. km.

Let us now return to our problem. First we calculate how much rain falls on each square kilometre of the earth's surface. The figure for 1 sq. m. or 10,000 sq. cm. is:

$$78 \times 10{,}000 = 780{,}000 \text{ cu. cm.}$$

A square kilometre contains $1{,}000 \times 1{,}000 = 1{,}000{,}000$ sq. m. The amount of rainfall for 1 sq. km. is therefore:

$$780{,}000{,}000{,}000 \text{ cu. cm. or } 780{,}000 \text{ cu. m.}$$

And for the entire surface of the globe the figure is:

$$780{,}000 \times 509{,}000{,}000 = 397{,}000{,}000{,}000{,}000 \text{ cu. m.}$$

To convert this into cubic kilometres it is necessary to divide it by $1{,}000 \times 1{,}000 \times 1{,}000$, i.e., by 1,000 million. The result is 397,000 cu. km.

Thus, the annual average amount of water falling on our earth from the atmosphere is equal to 400,000 cu. km. (in round figures).

Here we shall put a stop to our little talk about the geometry of rain and snow. We can get more detailed data about it in books on meteorology.

CHAPTER X

MATHEMATICS AND THE DELUGE

88. **THE DELUGE.**—The Bible tells us how the world was once flooded by rain-water that rose higher than the highest mountains. "It repented the Lord," it says, "that He had made man on the earth."

"I will destroy man whom I have created from the face of the earth," God said, "both man and beast, and the creeping thing, and the fowls of the air."

The only man God wanted to spare was Noah the just, and He warned him of the forthcoming destruction of the world and told him to build an ark 300 cubits in length, 50 cubits in breadth and 30 cubits in height. There were three storeys in the ark and it was to save not only Noah and his family and the families of his grown-up children, but also all the breeds of living creatures on earth. God ordered Noah to take two of every sort of these creatures into the ark with enough food for a lengthy period.

God chose deluge as a means to destroy all living things on earth. Water was to destroy all human beings and all beasts. After that Noah and the beasts he was to save would create a new human race and a new animal world.

"And it came to pass after seven days," the Bible continues, "that the waters of the flood were upon the earth. . . . And the rain was upon the earth forty days and forty nights. . . . And the waters increased, and bare up the ark. . . . And the waters prevailed exceedingly upon the earth; and all the high hills, that were under the whole heaven, were covered. Fifteen cubits upward did the waters prevail. . . . And all flesh died that moved upon the earth. . . . And Noah only remained alive, and they that were with him in the ark." The waters prevailed upon the earth, according to the Bible, another 110 days and then subsided, and Noah went forth out of the ark with all the creatures he had saved to replenish the earth.

The story of the deluge raises two questions:

1) Can any downpour cover the whole earth higher than the highest mountains?

2) Could Noah's ark really accommodate all the breeds of living creatures on earth?

89. WAS THE DELUGE POSSIBLE?—Both questions may be solved mathematically. Where did the deluge water come from? From the atmosphere, naturally. Where did it go after that? A whole world ocean of water could not be absorbed by soil, nor could it disappear in any other way. The only place where the waters could go was the atmosphere, i.e., they could evaporate. And that is where deluge water should be now. So, if all the vapour in the atmosphere were to condense into drops and fall on the earth, there would be another deluge with the water covering the highest mountains. Let us see if this is so.

Books on meteorology will tell us how much moisture there is in the atmosphere. They say that the column of air over each square metre contains on the average 16 kg. of vapour and never more than 25 kg. Let us calculate the depth of rain-water if all this vapour were to condense and fall on the earth. Twenty-five kilogrammes, i.e., 25,000 gr., of water would be equal in volume to 25,000 cu. cm. Such would be the volume of the stratum over an area of 1 cu. m., i.e., $100 \times 100 = 10,000$ sq. cm. Dividing the volume by the area of the base, we get the depth of the stratum:

$$25,000 : 10,000 = 2.5 \text{ cm.}$$

The flood-water could not have risen above 2.5 cm. because there would not have been enough water for that in the atmosphere.* And even this height could be possible only if the soil did not absorb rain-water.

Our calculation shows that the flood-water, even if there was a deluge, could not have risen higher than 2.5 cm. And it is a long way to the summit of Mount Everest, which is 9 km. high. The Bible exaggerated the height of the flood only . . . 360,000 times.

And so, even if there had been a rain "deluge," it could not have been a real one, but just a drizzle because 40 days of uninterrupted rain would have resulted in a precipitation of only 25 mm.—less than 0.5 mm. a day. An autumn drizzle that lasts a whole day brings 20 times more water.

90. WAS THERE SUCH AN ARK?—Let us now deal with the second question: could the ark accommodate all the creatures Noah had to save?

* In many places rainfall sometimes exceeds 2,5 cm., but in such cases it comes not only from the atmosphere directly over the given area, but is also brought by wind from the atmosphere of the neighbouring places. According to the Bible, the deluge *simultaneously* inundated the entire surface of the Earth and therefore one place could not "borrow" moisture from another.

Let us see how much space there was. According to the Bible, the ark was three storeys high. Each storey was 300 cubits long and 50 cubits wide. With the ancient West-Asian peoples a cubit was a measure of length approximating 45 cm. or 0.45 m. Converted into the metric system that means that each storey was

$$300 \times 0.45 = 135 \text{ m. long, and}$$
$$50 \times 0.45 = 22.5 \text{ m. wide.}$$

The area of each floor was therefore

$$135 \times 22.5 = 3{,}040 \text{ sq. m. (in round figure).}$$

And the total "living space" on all three floors was therefore:

$$3{,}040 \times 3 = 9{,}120 \text{ sq. m.}$$

Was that space enough, say, for just the *mammals*? There are approximately 3,500 kinds of mammals, and Noah had to allot space not only for the mammals themselves, but also for enough food to last the 150 days until the waters had completely subsided. Moreover, it should not be forgotten that beasts of prey required not only space for themselves, but also for their prey, as well as space for food for their prey. For each pair of mammals in the ark there were

$$9{,}120 : 3{,}500 = 2.6 \text{ sq. m. of space.}$$

That is definitely not enough, especially if we take into consideration that Noah and his large family also required some living space and that it was necessary to have intervals between cages.

Apart from the mammals, Noah had to take in many other living creatures, perhaps not so big as the mammals, but then far more variegated. Their number is something as follows:

Birds	13,000
Reptiles	3,500
Amphibians	1,400
Arachnids	16,000
Insects	360,000

If the mammals alone were cramped for space, then there was absolutely none for the other creatures. To accommodate everything living on earth the ark would have had to be very much bigger than it actually was. As it is, according to the Bible, the ark was a huge vessel—to use a seamen's expression, it was of 20,000 tons displacement. It is highly improbable that in those ancient days when the shipbuilding technique was in its infancy, people knew how to build vessels of such gigantic dimensions. But big as it was, it was not big enough to fulfil the task the Bible ascribes to it. The question was of a regular zoo with enough food for five months!

Briefly, the Bible story of the deluge is belied by mathematics. In fact, it is very unlikely there was anything of the sort. And if there was, it was probably some local flood—the rest was the work of the rich Eastern imagination.

CHAPTER XI
THIRTY DIFFERENT PROBLEMS

I hope that the reader has found this book quite useful, that it has not only entertained him, but also helped him to develop his wit and ingenuity and to make better use of his knowledge. The reader will

Fig. 89. Five sections of a chain.

no doubt like to test his ingenuity. To accommodate him I have worked out thirty different problems for the last chapter of the book.

91. A CHAIN.—A blacksmith was given a chain torn into five equal sections of three links each and was asked to fix it.

Before beginning the job, the blacksmith started to think how many links he would have to open up and then reforge. He finally decided on *four.*

Could this job be done with fewer links opened up and then reforged?

92. SPIDERS AND BEETLES.—A boy collected 8 spiders and beetles into a little box. He counted the legs and found there were altogether 54.

How many spiders and how many beetles did he collect?

93. A CAPE, A HAT AND GALOSHES.—A man bought a cape, a hat and a pair of galoshes and paid 140 rubles for the lot. The cape cost 90 rubles more than the hat, and the cape and the hat together 120 rubles more than the galoshes. How much did he pay for each item?

The problem should be solved mentally, without any equations.

94. CHICKEN AND DUCK EGGS.—The baskets (fig. 90) have eggs in them—chicken and duck. The number of eggs in each basket is written on the side. "If I sell this basket," the vendor says, "I shall have twice as many chicken eggs left as duck eggs."

Which basket did he have in mind?

Fig. 90. Which basket did he have in mind?

95. AN AIR TRIP.—A plane takes 1 hour and 20 minutes to fly from *A* to *B* and only 80 minutes to return. How do you explain that?

96. MONEY GIFTS.—Two fathers gave their two sons some money. One gave his son 150 rubles and the other 100 rubles. When the two sons counted their finances, they found that together they had become richer by only 150 rubles. What is the explanation?

97. TWO DRAUGHTS.—Place two draughts in any of the 64 squares of the board. How many different positions are there in which they can be arranged?

98. TWO DIGITS.—What is the smallest integer that can be written with two digits?

99. ONE.—Write 1 by using all the ten digits.

100. FIVE 9's.—Write 10 with five 9's. Do it in at least two ways.

101. TEN DIGITS.—Write 100 by using all the ten digits. How many ways are there of doing it? We know of at least four.

102. FOUR WAYS.—Show four different ways of writing 100 with five identical digits.

103. FOUR 1's.—What is the biggest number that can be written with four 1's?

104. MYSTERIOUS DIVISION.—In the following division all the digits except four 4's have been replaced by x's. Fill in the missing digits.

```
xxxxxx4 | xxx
xxx     | x4xx
-------
 xx4x
 xxxx
 ----
  xxxx
   x4x
   ----
   xxxx
   xxxx
   ----
```

There are several ways of solving this problem.

105. ANOTHER DIVISION.—Here is another division of this type, only now you have seven 7's to start with.

```
xx7xxxxxxx |xxxx7x
xxxxxx     |xx7xx
----------
xxxxx7x
xxxxxxx
-------
  x7xxxx
  x7xxxx
  ------
  xxxxxxx
  xxxx7xx
  -------
     xxxxxx
     xxxxxx
     ------
```

106. WHAT WILL YOU HAVE?—Calculate mentally the length of the strip of all the millimetre squares in one square metre, if placed one alongside the other.

107. SOMETHING OF THE SAME SORT.—Calculate mentally the height of a pole made up of all the millimetre cubes in one cubic metre, if placed one atop another.

108. A PLANE.—A plane with a 12-metre span was photographed as it was flying directly overhead. The depth of the camera is 12 cm. In the photo the plane had an 8-mm. span.

How high was the plane when it was snapped?

109. A MILLION THINGS.—A thing weighs 89.4 gr. Calculate mentally the weight in tons of one million of them.

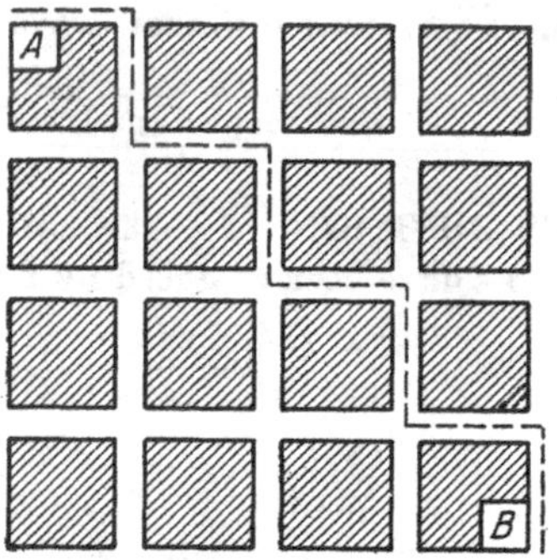

Fig. 91. The paths dividing the summer estate.

110. THE NUMBER OF WAYS.—Fig. 91 is a plan of a summer estate divided by paths into square lots. The dotted line is the way taken by a man to go from point *A* to point *B*.

This, of course, is not the only way between the two points. How many different ways of the same length are there?

111. THE CLOCK FACE.—You are to cut the face of the clock in fig. 92 into six parts of any shape, but the aggregate number in all of them must be the same.

The problem is a test for your resourcefulness and ingenuity.

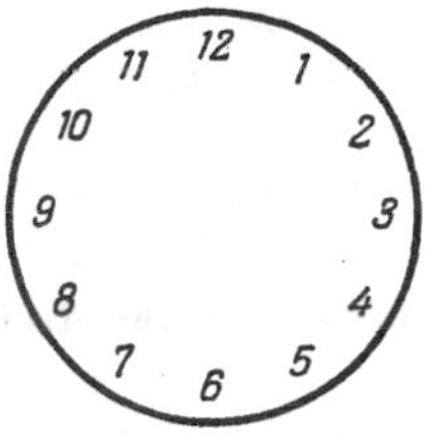

Fig. 92. Cut the clock into six face parts.

112. EIGHT-POINTED STAR.—In fig. 93 fill the little circles at the points of intersection with numbers from 1 to 16 in such a way as to have a total of 34 on each side of the square and 34 at the vertices.

113. A NUMBER WHEEL.—In fig. 94 arrange the numbers 1 to 9 in such a way as to have one of them in the centre and the rest at the ends of the diameters—the sum of the three numbers on each diameter to be 15.

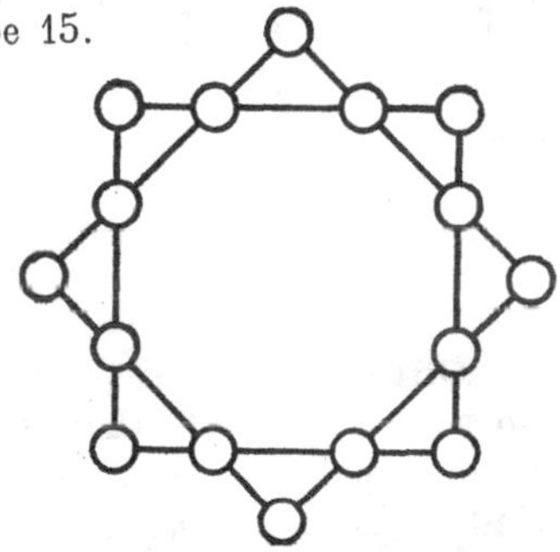
Fig. 93. The eight-pointed star.

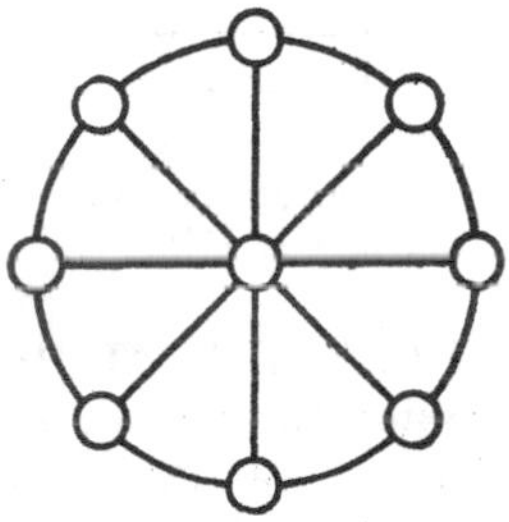
Fig. 94. The number wheel.

114. A TRIPOD.—It is claimed that a tripod always stands firmly, even when its three legs are of different length. Is that right?

115. ANGLES.—How many degrees are there in the angles formed by clock hands in fig. 95. The problem should be solved mentally, without a protractor.

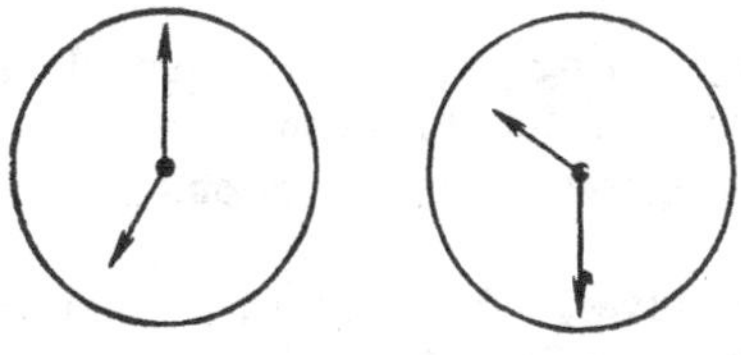
Fig. 95. How big are the angles?

116. ON THE EQUATOR.—If we could walk around the Earth on the equator, the top of our head would describe a circle whose circumference would be longer than the circle described by our feet.

How great is the difference?

117. SIX ROWS.—You have probably heard the joke of how nine horses were placed in ten stalls? Well, here is a problem that outwardly is very much alike, except that it can really be solved. Here it is:

Line up 24 persons in 6 rows, with each row to comprise 5.

118. HOW TO DIVIDE THAT?—There is a well-known problem calling for the division of the figure 96—a rectangle with a quarter of its area cut off—into four equal parts. What you are asked to do is to divide it into three equal parts. Is that possible?

119. THE CROSS AND THE CRESCENT.—In fig. 97 you see the figure of a crescent. The problem is to draw a red cross whose area geometrically would be equal to the area of the crescent.

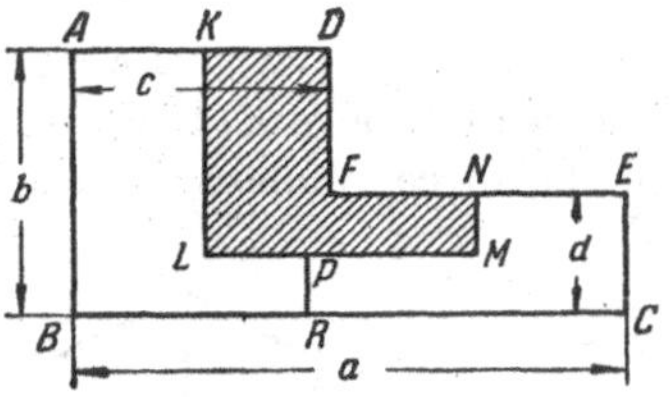

Fig. 96. Can you divide this into three equal parts?

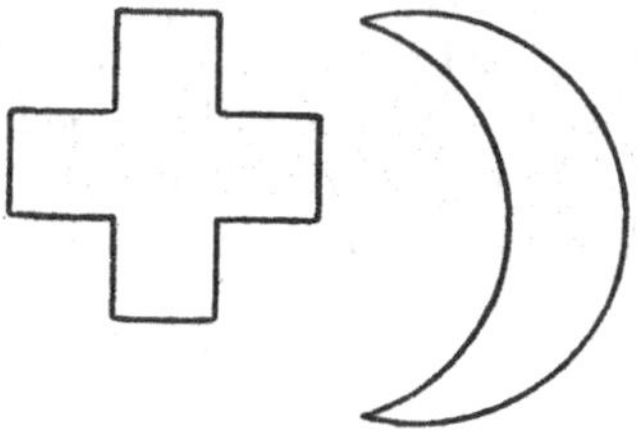

Fig. 97. How to make a cross out of a crescent?

120. THE BENEDIKTOV PROBLEM. —A great many lovers of Russian literature probably do not even suspect that the poet Benediktov (1807-1873) collected and compiled a whole volume of mathematical conundrums. Had it been published, it would have been the first Russian book of this type. But it never was and the manuscript was found only in 1924. I had the good fortune to study the manuscript and even established—by solving one of the brainteasers contained therein—that the collection was completed in 1869 (the manuscript itself was not dated). The following problem in the form of a story is given below. It is captioned: "An Ingenious Way of Solving a Tricky Problem."

"One woman who made a living by selling eggs had 90 eggs which she wanted sold, so she sent her three daughters to the market, giving 10 eggs to her eldest and cleverest daughter, 30 to the second and 50 to the third.

" 'You'd better agree among yourselves,' she told them, 'about the price you're going to ask for the eggs, and keep to it. Stick to the price you decide upon. But I hope that, in spite of your agreement, the eldest, being the cleverest, will receive as much for her ten eggs as the second will receive for her thirty and will teach the second to sell her thirty eggs for as much as the youngest will sell her fifty. In other words, each of you is to bring home the same amount, and the total for the 90 eggs is not to be less than 90 kopeks.'"

I shall stop here to allow the reader to rack his brains a little and think how the girls carried out the instructions.

Answers 91 to 120

91. The job can be done by opening up just *three* links, that is, the links of one section, and joining the ends of the other four sections with them.

92. Before you solve this problem you must know how many legs the spider and the beetle have. If you remember your natural science, you know that spiders have 8 legs and beetles 6.

Now let us assume that there were only beetles—8 of them—in the box. That means there should be $6\times8=48$ legs, or 6 less than mentioned in the problem. If we substitute one spider for one of the beetles, the number of legs will increase by 2 because the spider has 8 legs and not 6.

It is clear that if we substitute three spiders for three beetles we shall bring the number of legs in the box to the required 54. Then instead of 8 beetles we shall have 5, the rest will be spiders.

Hence, the boy collected 5 beetles and 3 spiders.

Let us verify: 5 beetles have 30 legs and 3 spiders have 24 legs. And $30+24=54$.

There is another way of solving the problem. We may assume that there were only spiders in the box—8 of them. Then we would have $8 \times 8 = 64$ legs, i.e., 10 more than stated in the problem. By substituting one beetle for one spider we shall decrease the number of legs by 2. On the whole, we require 5 such substitutions to bring the number of legs to the required 54. In other words, we must leave 3 of the 8 spiders in the box and replace the others with beetles.

93. If instead of a cape, a hat and galoshes he were to buy only two pairs of galoshes, he would have to pay not 140 rubles but as many times less as the galoshes are cheaper than the cape and the hat, i.e., 120 rubles less. Consequently, the two pairs of galoshes cost $140 - 120 = 20$ rubles. Hence, one pair costs 10 rubles.

Now we know that the cape and the hat together cost $140 - 10 = 130$ rubles. We also know that the cape is 90 rubles dearer than the hat. Let us use the same argument: let us buy two hats instead of a cape and a hat. In that case we should have to pay not 130 rubles, but 90 rubles less. Therefore, two hats cost $130 - 90 = 40$ rubles, and one hat 20 rubles.

So, here is how much each article would cost: galoshes—10 rubles, hat—20 rubles and cape—110 rubles.

94. The egg vendor had in mind the basket with 29 eggs. Chicken eggs were in the baskets marked 23, 12 and 5; duck eggs in the baskets marked 14 and 6.

Let us verify the answer. There would be

$23 + 12 + 5 = 40$ chicken eggs after the sale, and
$14 + 6 = 20$ duck eggs.

Thus, as the problem said, there were twice as many chicken eggs as duck eggs.

95. There is nothing to explain here. The plane made the flights there and back in absolutely the same time because 80 minutes and 1 hour and 20 minutes are one and the same thing.

The problem is meant for an inattentive reader who may think that there is some difference between 80 minutes and 1 hour 20 minutes.

96. The whole trick is that one of the fathers is the son of the other father. There are only three persons in this problem and not 4: grandfather, father and son. Grandfather gave his son 150 rubles and the latter passed on 100 of them to the grandson (i.e., his son), thus increasing his own capital by 50 rubles.

97. The first piece can be placed in any of the 64 squares, i.e., there are 64 ways of placing it. When it has been placed, there remain 63 squares for the second piece. In other words, to any one of the 64 positions of the first piece we can add the 63 positions of the second.

Hence, there are $64 \times 63 = 4{,}032$ different positions in which two pieces may be placed on a draughtboard.

98. The smallest integer that can be written with two digits is not 10, as some might think, but 1, expressed as follows:

$$\frac{1}{1}, \frac{2}{2}, \frac{3}{3}, \frac{4}{4}, \text{ etc., up to } \frac{9}{9}.$$

People acquainted with algebra can add a number of other expressions:

$$1^{0}, 2^{0}, 3^{0}, 4^{0}, \text{ etc., up to } 9^{0},$$

because any number raised to the zero power is equal to one.*

99. It is necessary to present 1 as a sum of two fractions:

$$\frac{148}{296} + \frac{35}{70} = 1$$

Those who know algebra have other answers:

$$123456789^{0};\quad 234567^{9-8-1}, \text{ etc.}$$

100. The two ways are as follows:

$$9\frac{99}{99} = 10, \text{ and}$$

$$\frac{99}{9} - \frac{9}{9} = 10$$

If you know algebra, you will probably add several other solutions. For instance:

$$\left(9\frac{9}{9}\right)^{\frac{9}{9}} = 10$$

$$9 + 99^{9-9} = 10$$

101. Here are four solutions:

$$70 + 24\frac{9}{18} + 5\frac{3}{6} = 100;$$

$$80\frac{27}{54} + 19\frac{3}{6} = 100;$$

$$87 + 9\frac{4}{5} + 3\frac{12}{60} = 100;$$

$$50\frac{1}{2} + 49\frac{38}{76} = 100$$

102. It is easy to write 100 by using five identical digits—1's, 3's or—the simplest way—5's. Here we see the four ways:

$$111 - 11 = 100;$$

$$33 \times 3 + \frac{3}{3} = 100;$$

$$5 \times 5 \times 5 - 5 \times 5 = 100;$$

$$(5+5+5+5) \times 5 = 100$$

103. People often say that it is 1,111. But it is possible to write a number many, many times greater, namely 11^{11}, i.e., the eleventh power of 11. If you have enough patience to calculate it to the end (with logarithms such a process can be much simplified), you will

---

* It would be wrong to write $\frac{0}{0}$ or 0^{0}; these expressions have no sense.

find that the total exceeds 280,000,000,000. Therefore, it is 250 million times greater than 1,111.

104. This problem may be solved in four different ways, namely:

$$1{,}337{,}174 : 943 = 1{,}418;$$
$$1{,}343{,}784 : 949 = 1{,}416;$$
$$1{,}200{,}474 : 846 = 1{,}419;$$
$$1{,}202{,}464 : 848 = 1{,}418$$

105. This problem has only one solution:

$$7{,}375{,}428{,}413 : 125{,}473 = 58{,}781.*$$

These two last, and rather difficult, problems were first published in the American publications *School World* (1906) and *Mathematical Magazine* (1920).

106. A square metre equals 1,000 thousand square millimetres. One thousand millimetre squares placed one alongside the other will stretch out 1 metre; 1,000 thousand squares will therefore be 1,000 metres long, i.e., 1 kilometre long.

107. The answer is stunning: the pole would be . . . 1,000 kilometres high! Let us calculate it mentally. A cubic metre is equal to 1,000 cubic millimetre × 1,000 × 1,000. One thousand millimetre cubes placed one atop another will make a pole 1 metre in height. And since we have 1,000 × 1,000 times more cubes, we shall have a pole that is 1,000 kilometres long.

Fig. 98.

108. Fig. 98 shows that (since angles 1 and 2 are equal) the linear measurements of the object are to the corresponding measurements of the picture as the distance of the object from the lens is to the depth of the camera. In our case, having taken x as the height of the plane (in metres), we come to the following proportion:

$$12{,}000 : 8 = x : 0{,}12.$$

Hence, $x = 180$ metres.

109. Here is how this calculation is done mentally. You must multiply 89.4 gr. by one million, i.e., by 1,000 thousands.

* Three other solutions were discovered later.

We do it in two stages: 89.4 gr. ×1,000=89.4 kg., because a kilogramme is 1,000 times more than a gramme. Then, 89.4 kg. ×1,000= 89.4 tons because a ton is 1,000 times more than a kilogramme.

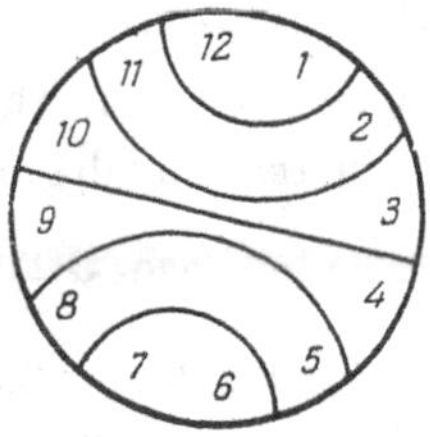

Fig. 99.

So, the weight we seek is 89.4 tons.

110. There are altogether 70 ways of going from *A* to *B*. (Systematic solution of this problem is possible with the aid of Pascal's triangle, as studied in algebra.)

111. As the total of all the numbers on a clock face is 78, the total of each of the six parts should be 78 : 6=13. This helps to find the solution (which is shown in fig. 99).

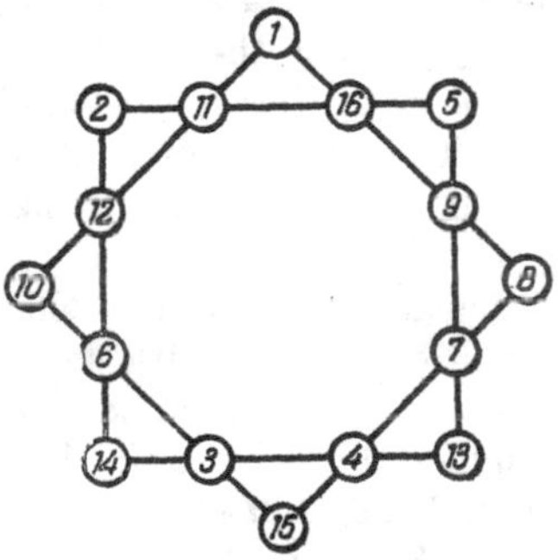

Fig. 100.

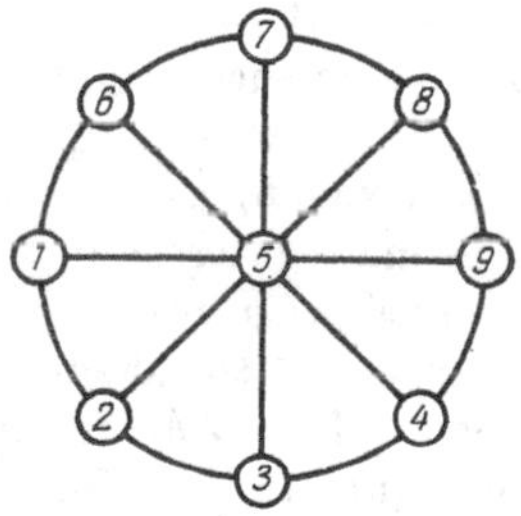

Fig. 101.

112. and **113.** The solutions are given in figures 100 and 101.

114. The three legs of a tripod always rest on the floor because through any three points of space there can pass one plane, and only one. That is the reason why the tripod stands firmly. As you see, the reason is purely geometrical and not physical.

That is why tripods are so convenient for land-surveying instruments and photographic cameras. A fourth leg would not make it any firmer; on the contrary, it would only be a bother.

115. The problem is easy to answer, especially if you see what time it is. The hands on the clock seen on the left (in fig. 95) show that it is 7 o'clock. This means that the arc between the two numbers is equal to $\frac{5}{12}$ of the circumference.

In terms of degrees this is equal to:

$$360° \times \frac{5}{12} = 150°.$$

The hands on the right show 9.30 o'clock. The arc here is equal to $3\frac{1}{2}$, or $\frac{7}{24}$, of the circumference.

In terms of degrees this is equal to

$$360° \times \frac{7}{24} = 105°.$$

Fig. 102.

116. If we assume that the average height of man is 175 cm. and take R as the radius of the Earth, we have $2\times 3.14\times(R+175)-(2\times 3.14\times R)=2\times 3.14\times 175=1{,}100$ cm., i.e., 11 metres. The surprising thing is that the result does in no way depend on the radius of the globe and, therefore, would be the same whether on such a gigantic planet as the Sun or a little ball.

117. It is easy to solve this problem if we arrange the men in the form of a hexagon, as shown in fig. 102.

118. The main interest in this problem lies in the fact that it cannot be solved with any a, b, c, d and e, but just with definite ones.

Indeed, we want the shaded part in fig. 96 to equal each of the unshaded parts. Line LM is obviously shorter than BC; therefore, it should be equal to AB. On the other hand, LM should be equal to RC; hence, $LM=RC=b$. Therefore, $BR=a-b$. But BR should be equal to KL and CE. That means $BR=KL=CE$, i.e., $a-b=d$ and $KL=d$.

We see that a, b and d cannot be chosen arbitrarily. Side d should be equal to the difference between sides a and b. But that is not enough. We shall see now that all the sides must constitute definite portions of side a.

We evidently have $PR+KL=AB$ or $PR+(a-b)=b$, i. e., $PR=2b-a$. Comparing the corresponding sides of the shaded part and the unshaded one on the right, we get: $PR=MN$, i.e., $PR=\frac{d}{2}$; hence $\frac{d}{2}=2b-a$. Comparing the last equation with the relation $a-b=d$, we find: $b=\frac{3}{5}a$ and $d=\frac{2}{5}a$. Comparison of the shaded part and the unshaded one on the left shows us that $AK=MN$, i.e.,

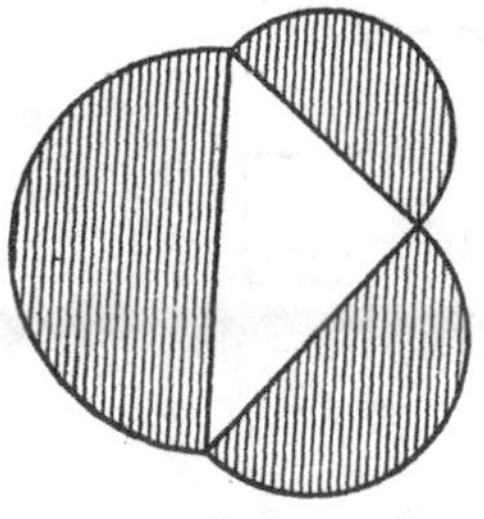

Fig. 103.

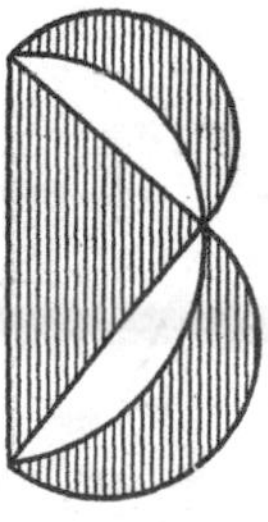

Fig. 104.

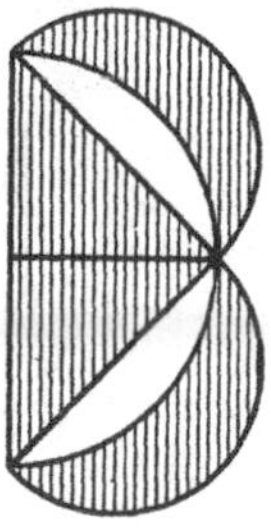

Fig. 105.

$AK=PR=\frac{d}{2}=\frac{1}{5}a$. Thus, we see that $KD=PR=\frac{1}{5}a$. Therefore, $AD=\frac{2}{5}a$.

And so, the sides of our figure cannot be taken arbitrarily. They must be definite portions $\left(\frac{3}{5}, \frac{2}{5} \text{ and } \frac{2}{5}\right)$ of side a. Solution is possible only in this case.

119. Readers who have heard that it is impossible to square the circle will probably think that geometrically this problem is also insoluble. If you can't square a circle, many think, how can you turn a crescent—which is formed of two arcs—into a rectangular figure?

Nevertheless, the problem can definitely be solved by geometrical construction through the application of one of the interesting corollaries of the well-known Pythagorean Proposition: that the semicircle formed on the hypotenuse is equal to the sum of the semicircles formed on the two other sides (fig. 103). Throwing the big semicircle over to the other side (fig. 104), we see that the two shaded crescents taken together are equal to the triangle.* If we take an isosceles triangle, then each of these two crescents will be equal to half of this triangle (fig. 105).

Hence, geometrically it is possible to form a right isosceles triangle whose area will be equal to that of a crescent.

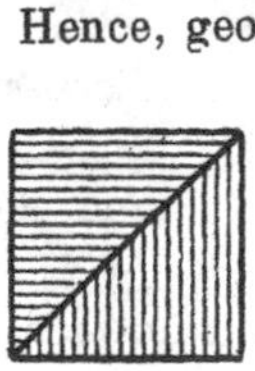

Fig. 106.

And since a right isosceles triangle can easily be converted into a square (fig. 106), it is possible geometrically to replace our crescent with a square.

There now remains only to turn this square into an equivalent cross (comprising, as it is known, 5 equal-sized squares). There are several ways of doing that: two of them are shown in figures 107 and 108, and both begin with uniting the vertices of the square with the centres of the opposite sides.

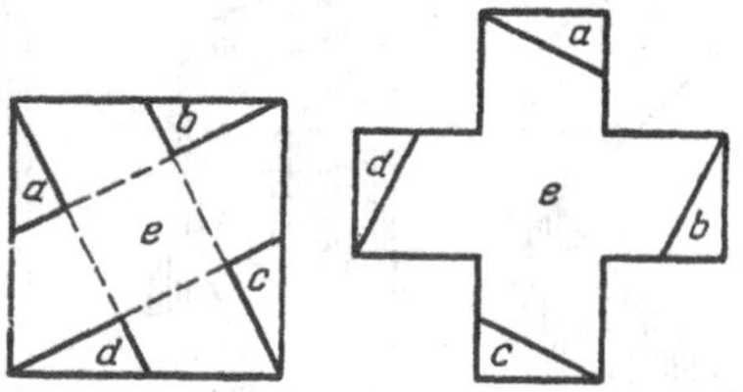

Fig. 107.

It should be borne in mind, however, that to turn a crescent into a cross of similar area we must have a crescent formed of two arcs of the circumference: the external arc, or a semicircle, and the internal arc, or a quarter of the circumference of a relatively bigger radius.**

I have shown you how to construct a cross equal in area to a crescent. The ends *A* and *B* of the crescent (fig. 109) are connected by a straight line; in the centre *O* of this straight line we raise a perpendicular and lay off $OC=O4$. The isosceles triangle *OAC* is then filled out to form the square *OADC* which is turned into a cross by one of the ways shown in figures 107 and 108.

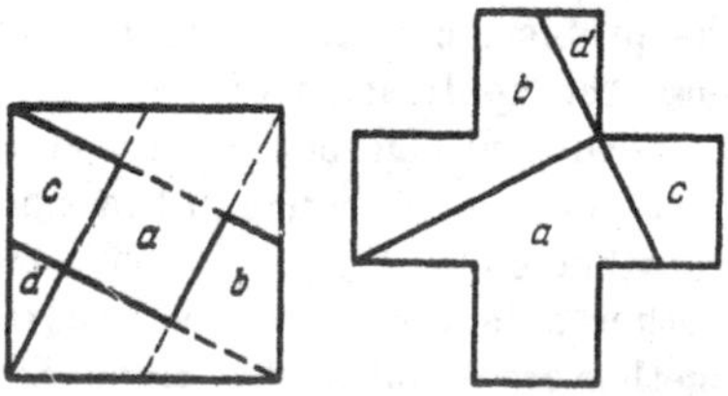

Fig. 108.

121. Let us continue with the Benediktov story:

"The problem was a tricky one and the three daughters discussed it on their way to the market, with the second and third appealing to the eldest for advice. The latter thought for a while and then said:

"'Look, sisters, we'll sell our eggs seven at a time and not ten as we always do. We'll fix a price for seven eggs and stick to it, as mother has told us to. Mind you, don't reduce the price however much

---

* In geometry this relation is known as the "Hippocrates's lunes."

** The crescent we see in the sky is somewhat dieffrent in form: its external arc is a semicircle and the internal is a semiellipse. Artists often paint it wrong by forming it of arcs of circumferences.

people may bargain! We'll ask three kopeks for the first seven eggs, all right?'

"'That's pretty cheap!' the second sister interjected.

"'Never mind,' the eldest retorted, 'we'll raise the price for the eggs that remain after that. I have made sure that there won't be any other egg vendors at the market. So there'll be no one to force our prices down. And when there's a demand for eggs and not many of them are left, the price goes up, that's only natural. And that's exactly where we'll make up.'

"'And how much shall we ask for the remaining eggs?' the youngest sister asked.

"'Nine kopeks an egg. And believe me, people who need eggs will pay the price.'

"'That's pretty stiff,' the second sister remarked.

"'So what? The first seven-egg batches will be cheap. The expensive eggs will make up for the loss.'

"The sisters agreed.

"At the market each chose a place. The cheap price brought on an avalanche of buyers and the youngest, who had 50 eggs, soon sold all her eggs but one. At three kopeks per seven eggs she made 21 kopeks. The second sister, who had 30 eggs, made 12 kopeks by selling four people seven eggs each, and had two eggs left in the basket. The eldest made 3 kopeks from the sale of seven eggs and was left with three eggs.

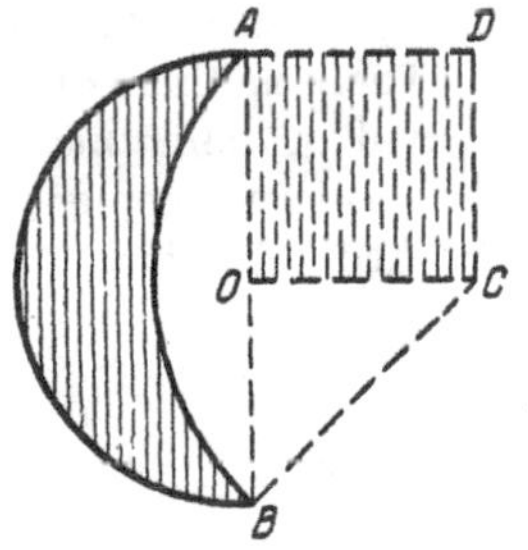

Fig. 109.

"Suddenly a cook appeared with instructions to buy ten eggs. Her mistress's sons had come home on leave and they loved an omelette. The cook rushed about the market, but the only vendors were the three sisters and then they had only six eggs—the youngest had one, the second sister two and the eldest three.

"It is only natural that the cook rushed to the one who had three—that is, to the eldest sister who had sold her batch of seven eggs for 3 kopeks.

"'How much d'you want for your eggs?' she asked.

"'Nine kopeks an egg,' was the reply.

"'What!? You're crazy!'

"'Take them or leave them. These are my last and I won't take a kopek less.'

"The cook ran to the second sister, the one who had two eggs left in her basket.

"'How much?' she yelled.

"'Nine kopeks an egg. That's the price and these are my last eggs.'

"'And how much do you want for your egg?' the cook turned to the youngest sister.

"'Nine kopeks.'

"Well, there was nothing the cook could do, so she bought the eggs at this exorbitant price.

"'All right,' she burst out, 'I'll take the lot.'

"She paid 27 kopeks to the eldest sister for her three eggs and with the three kopeks the latter had from the earlier sale this brought her total receipts to 30 kopeks. The second sister got 18 kopeks and with the 12 kopeks she had received earlier that also made 30 kopeks. The youngest got 9 kopeks for the remaining egg and that, added to the 21 kopeks she had made on the sale of 49 eggs, brought the total also to 30 kopeks.

"The three sisters then returned home, gave the money to their mother and told her how, sticking to the price they had agreed upon, they had succeeded in selling ten eggs for the same price as 50.

"Their mother was very pleased that her instructions had been carried out and that her eldest daughter had proved so clever. But she was even happier that her daughters had brought her exactly what she had told them to bring—90 kopeks."

A CATALOG OF SELECTED

DOVER BOOKS

IN SCIENCE AND MATHEMATICS

Astronomy

CHARIOTS FOR APOLLO: The NASA History of Manned Lunar Spacecraft to 1969, Courtney G. Brooks, James M. Grimwood, and Loyd S. Swenson, Jr. This illustrated history by a trio of experts is the definitive reference on the Apollo spacecraft and lunar modules. It traces the vehicles' design, development, and operation in space. More than 100 photographs and illustrations. 576pp. 6 3/4 x 9 1/4. 0-486-46756-2

EXPLORING THE MOON THROUGH BINOCULARS AND SMALL TELESCOPES, Ernest H. Cherrington, Jr. Informative, profusely illustrated guide to locating and identifying craters, rills, seas, mountains, other lunar features. Newly revised and updated with special section of new photos. Over 100 photos and diagrams. 240pp. 8 1/4 x 11. 0-486-24491-1

WHERE NO MAN HAS GONE BEFORE: A History of NASA's Apollo Lunar Expeditions, William David Compton. Introduction by Paul Dickson. This official NASA history traces behind-the-scenes conflicts and cooperation between scientists and engineers. The first half concerns preparations for the Moon landings, and the second half documents the flights that followed Apollo 11. 1989 edition. 432pp. 7 x 10. 0-486-47888-2

APOLLO EXPEDITIONS TO THE MOON: The NASA History, Edited by Edgar M. Cortright. Official NASA publication marks the 40th anniversary of the first lunar landing and features essays by project participants recalling engineering and administrative challenges. Accessible, jargon-free accounts, highlighted by numerous illustrations. 336pp. 8 3/8 x 10 7/8. 0-486-47175-6

ON MARS: Exploration of the Red Planet, 1958-1978--The NASA History, Edward Clinton Ezell and Linda Neuman Ezell. NASA's official history chronicles the start of our explorations of our planetary neighbor. It recounts cooperation among government, industry, and academia, and it features dozens of photos from Viking cameras. 560pp. 6 3/4 x 9 1/4. 0-486-46757-0

ARISTARCHUS OF SAMOS: The Ancient Copernicus, Sir Thomas Heath. Heath's history of astronomy ranges from Homer and Hesiod to Aristarchus and includes quotes from numerous thinkers, compilers, and scholasticists from Thales and Anaximander through Pythagoras, Plato, Aristotle, and Heraclides. 34 figures. 448pp. 5 3/8 x 8 1/2. 0-486-43886-4

AN INTRODUCTION TO CELESTIAL MECHANICS, Forest Ray Moulton. Classic text still unsurpassed in presentation of fundamental principles. Covers rectilinear motion, central forces, problems of two and three bodies, much more. Includes over 200 problems, some with answers. 437pp. 5 3/8 x 8 1/2. 0-486-64687-4

BEYOND THE ATMOSPHERE: Early Years of Space Science, Homer E. Newell. This exciting survey is the work of a top NASA administrator who chronicles technological advances, the relationship of space science to general science, and the space program's social, political, and economic contexts. 528pp. 6 3/4 x 9 1/4. 0-486-47464-X

STAR LORE: Myths, Legends, and Facts, William Tyler Olcott. Captivating retellings of the origins and histories of ancient star groups include Pegasus, Ursa Major, Pleiades, signs of the zodiac, and other constellations. "Classic." – *Sky & Telescope.* 58 illustrations. 544pp. 5 3/8 x 8 1/2. 0-486-43581-4

A COMPLETE MANUAL OF AMATEUR ASTRONOMY: Tools and Techniques for Astronomical Observations, P. Clay Sherrod with Thomas L. Koed. Concise, highly readable book discusses the selection, set-up, and maintenance of a telescope; amateur studies of the sun; lunar topography and occultations; and more. 124 figures. 26 halftones. 37 tables. 335pp. 6 1/2 x 9 1/4. 0-486-42820-6

Chemistry

MOLECULAR COLLISION THEORY, M. S. Child. This high-level monograph offers an analytical treatment of classical scattering by a central force, quantum scattering by a central force, elastic scattering phase shifts, and semi-classical elastic scattering. 1974 edition. 310pp. 5 3/8 x 8 1/2. 0-486-69437-2

HANDBOOK OF COMPUTATIONAL QUANTUM CHEMISTRY, David B. Cook. This comprehensive text provides upper-level undergraduates and graduate students with an accessible introduction to the implementation of quantum ideas in molecular modeling, exploring practical applications alongside theoretical explanations. 1998 edition. 832pp. 5 3/8 x 8 1/2. 0-486-44307-8

RADIOACTIVE SUBSTANCES, Marie Curie. The celebrated scientist's thesis, which directly preceded her 1903 Nobel Prize, discusses establishing atomic character of radioactivity; extraction from pitchblende of polonium and radium; isolation of pure radium chloride; more. 96pp. 5 3/8 x 8 1/2. 0-486-42550-9

CHEMICAL MAGIC, Leonard A. Ford. Classic guide provides intriguing entertainment while elucidating sound scientific principles, with more than 100 unusual stunts: cold fire, dust explosions, a nylon rope trick, a disappearing beaker, much more. 128pp. 5 3/8 x 8 1/2. 0-486-67628-5

ALCHEMY, E. J. Holmyard. Classic study by noted authority covers 2,000 years of alchemical history: religious, mystical overtones; apparatus; signs, symbols, and secret terms; advent of scientific method, much more. Illustrated. 320pp. 5 3/8 x 8 1/2. 0-486-26298-7

CHEMICAL KINETICS AND REACTION DYNAMICS, Paul L. Houston. This text teaches the principles underlying modern chemical kinetics in a clear, direct fashion, using several examples to enhance basic understanding. Solutions to selected problems. 2001 edition. 352pp. 8 3/8 x 11. 0-486-45334-0

PROBLEMS AND SOLUTIONS IN QUANTUM CHEMISTRY AND PHYSICS, Charles S. Johnson and Lee G. Pedersen. Unusually varied problems, with detailed solutions, cover of quantum mechanics, wave mechanics, angular momentum, molecular spectroscopy, scattering theory, more. 280 problems, plus 139 supplementary exercises. 430pp. 6 1/2 x 9 1/4. 0-486-65236-X

ELEMENTS OF CHEMISTRY, Antoine Lavoisier. Monumental classic by the founder of modern chemistry features first explicit statement of law of conservation of matter in chemical change, and more. Facsimile reprint of original (1790) Kerr translation. 539pp. 5 3/8 x 8 1/2. 0-486-64624-6

MAGNETISM AND TRANSITION METAL COMPLEXES, F. E. Mabbs and D. J. Machin. A detailed view of the calculation methods involved in the magnetic properties of transition metal complexes, this volume offers sufficient background for original work in the field. 1973 edition. 240pp. 5 3/8 x 8 1/2. 0-486-46284-6

GENERAL CHEMISTRY, Linus Pauling. Revised third edition of classic first-year text by Nobel laureate. Atomic and molecular structure, quantum mechanics, statistical mechanics, thermodynamics correlated with descriptive chemistry. Problems. 992pp. 5 3/8 x 8 1/2. 0-486-65622-5

ELECTROLYTE SOLUTIONS: Second Revised Edition, R. A. Robinson and R. H. Stokes. Classic text deals primarily with measurement, interpretation of conductance, chemical potential, and diffusion in electrolyte solutions. Detailed theoretical interpretations, plus extensive tables of thermodynamic and transport properties. 1970 edition. 590pp. 5 3/8 x 8 1/2. 0-486-42225-9

Engineering

FUNDAMENTALS OF ASTRODYNAMICS, Roger R. Bate, Donald D. Mueller, and Jerry E. White. Teaching text developed by U.S. Air Force Academy develops the basic two-body and n-body equations of motion; orbit determination; classical orbital elements, coordinate transformations; differential correction; more. 1971 edition. 455pp. 5 3/8 x 8 1/2. 0-486-60061-0

INTRODUCTION TO CONTINUUM MECHANICS FOR ENGINEERS: Revised Edition, Ray M. Bowen. This self-contained text introduces classical continuum models within a modern framework. Its numerous exercises illustrate the governing principles, linearizations, and other approximations that constitute classical continuum models. 2007 edition. 320pp. 6 1/8 x 9 1/4. 0-486-47460-7

ENGINEERING MECHANICS FOR STRUCTURES, Louis L. Bucciarelli. This text explores the mechanics of solids and statics as well as the strength of materials and elasticity theory. Its many design exercises encourage creative initiative and systems thinking. 2009 edition. 320pp. 6 1/8 x 9 1/4. 0-486-46855-0

FEEDBACK CONTROL THEORY, John C. Doyle, Bruce A. Francis and Allen R. Tannenbaum. This excellent introduction to feedback control system design offers a theoretical approach that captures the essential issues and can be applied to a wide range of practical problems. 1992 edition. 224pp. 6 1/2 x 9 1/4. 0-486-46933-6

THE FORCES OF MATTER, Michael Faraday. These lectures by a famous inventor offer an easy-to-understand introduction to the interactions of the universe's physical forces. Six essays explore gravitation, cohesion, chemical affinity, heat, magnetism, and electricity. 1993 edition. 96pp. 5 3/8 x 8 1/2. 0-486-47482-8

DYNAMICS, Lawrence E. Goodman and William H. Warner. Beginning engineering text introduces calculus of vectors, particle motion, dynamics of particle systems and plane rigid bodies, technical applications in plane motions, and more. Exercises and answers in every chapter. 619pp. 5 3/8 x 8 1/2. 0-486-42006-X

ADAPTIVE FILTERING PREDICTION AND CONTROL, Graham C. Goodwin and Kwai Sang Sin. This unified survey focuses on linear discrete-time systems and explores natural extensions to nonlinear systems. It emphasizes discrete-time systems, summarizing theoretical and practical aspects of a large class of adaptive algorithms. 1984 edition. 560pp. 6 1/2 x 9 1/4. 0-486-46932-8

INDUCTANCE CALCULATIONS, Frederick W. Grover. This authoritative reference enables the design of virtually every type of inductor. It features a single simple formula for each type of inductor, together with tables containing essential numerical factors. 1946 edition. 304pp. 5 3/8 x 8 1/2. 0-486-47440-2

THERMODYNAMICS: Foundations and Applications, Elias P. Gyftopoulos and Gian Paolo Beretta. Designed by two MIT professors, this authoritative text discusses basic concepts and applications in detail, emphasizing generality, definitions, and logical consistency. More than 300 solved problems cover realistic energy systems and processes. 800pp. 6 1/8 x 9 1/4. 0-486-43932-1

THE FINITE ELEMENT METHOD: Linear Static and Dynamic Finite Element Analysis, Thomas J. R. Hughes. Text for students without in-depth mathematical training, this text includes a comprehensive presentation and analysis of algorithms of time-dependent phenomena plus beam, plate, and shell theories. Solution guide available upon request. 672pp. 6 1/2 x 9 1/4. 0-486-41181-8

HELICOPTER THEORY, Wayne Johnson. Monumental engineering text covers vertical flight, forward flight, performance, mathematics of rotating systems, rotary wing dynamics and aerodynamics, aeroelasticity, stability and control, stall, noise, and more. 189 illustrations. 1980 edition. 1089pp. 5 5/8 x 8 1/4. 0-486-68230-7

MATHEMATICAL HANDBOOK FOR SCIENTISTS AND ENGINEERS: Definitions, Theorems, and Formulas for Reference and Review, Granino A. Korn and Theresa M. Korn. Convenient access to information from every area of mathematics: Fourier transforms, Z transforms, linear and nonlinear programming, calculus of variations, random-process theory, special functions, combinatorial analysis, game theory, much more. 1152pp. 5 3/8 x 8 1/2. 0-486-41147-8

A HEAT TRANSFER TEXTBOOK: Fourth Edition, John H. Lienhard V and John H. Lienhard IV. This introduction to heat and mass transfer for engineering students features worked examples and end-of-chapter exercises. Worked examples and end-of-chapter exercises appear throughout the book, along with well-drawn, illuminating figures. 768pp. 7 x 9 1/4. 0-486-47931-5

BASIC ELECTRICITY, U.S. Bureau of Naval Personnel. Originally a training course; best nontechnical coverage. Topics include batteries, circuits, conductors, AC and DC, inductance and capacitance, generators, motors, transformers, amplifiers, etc. Many questions with answers. 349 illustrations. 1969 edition. 448pp. 6 1/2 x 9 1/4. 0-486-20973-3

BASIC ELECTRONICS, U.S. Bureau of Naval Personnel. Clear, well-illustrated introduction to electronic equipment covers numerous essential topics: electron tubes, semiconductors, electronic power supplies, tuned circuits, amplifiers, receivers, ranging and navigation systems, computers, antennas, more. 560 illustrations. 567pp. 6 1/2 x 9 1/4. 0-486-21076-6

BASIC WING AND AIRFOIL THEORY, Alan Pope. This self-contained treatment by a pioneer in the study of wind effects covers flow functions, airfoil construction and pressure distribution, finite and monoplane wings, and many other subjects. 1951 edition. 320pp. 5 3/8 x 8 1/2. 0-486-47188-8

SYNTHETIC FUELS, Ronald F. Probstein and R. Edwin Hicks. This unified presentation examines the methods and processes for converting coal, oil, shale, tar sands, and various forms of biomass into liquid, gaseous, and clean solid fuels. 1982 edition. 512pp. 6 1/8 x 9 1/4. 0-486-44977-7

THEORY OF ELASTIC STABILITY, Stephen P. Timoshenko and James M. Gere. Written by world-renowned authorities on mechanics, this classic ranges from theoretical explanations of 2- and 3-D stress and strain to practical applications such as torsion, bending, and thermal stress. 1961 edition. 560pp. 5 3/8 x 8 1/2. 0-486-47207-8

PRINCIPLES OF DIGITAL COMMUNICATION AND CODING, Andrew J. Viterbi and Jim K. Omura. This classic by two digital communications experts is geared toward students of communications theory and to designers of channels, links, terminals, modems, or networks used to transmit and receive digital messages. 1979 edition. 576pp. 6 1/8 x 9 1/4. 0-486-46901-8

LINEAR SYSTEM THEORY: The State Space Approach, Lotfi A. Zadeh and Charles A. Desoer. Written by two pioneers in the field, this exploration of the state space approach focuses on problems of stability and control, plus connections between this approach and classical techniques. 1963 edition. 656pp. 6 1/8 x 9 1/4. 0-486-46663-9

Mathematics–Bestsellers

HANDBOOK OF MATHEMATICAL FUNCTIONS: with Formulas, Graphs, and Mathematical Tables, Edited by Milton Abramowitz and Irene A. Stegun. A classic resource for working with special functions, standard trig, and exponential logarithmic definitions and extensions, it features 29 sets of tables, some to as high as 20 places. 1046pp. 8 x 10 1/2. 0-486-61272-4

ABSTRACT AND CONCRETE CATEGORIES: The Joy of Cats, Jiri Adamek, Horst Herrlich, and George E. Strecker. This up-to-date introductory treatment employs category theory to explore the theory of structures. Its unique approach stresses concrete categories and presents a systematic view of factorization structures. Numerous examples. 1990 edition, updated 2004. 528pp. 6 1/8 x 9 1/4. 0-486-46934-4

MATHEMATICS: Its Content, Methods and Meaning, A. D. Aleksandrov, A. N. Kolmogorov, and M. A. Lavrent'ev. Major survey offers comprehensive, coherent discussions of analytic geometry, algebra, differential equations, calculus of variations, functions of a complex variable, prime numbers, linear and non-Euclidean geometry, topology, functional analysis, more. 1963 edition. 1120pp. 5 3/8 x 8 1/2. 0-486-40916-3

INTRODUCTION TO VECTORS AND TENSORS: Second Edition--Two Volumes Bound as One, Ray M. Bowen and C.-C. Wang. Convenient single-volume compilation of two texts offers both introduction and in-depth survey. Geared toward engineering and science students rather than mathematicians, it focuses on physics and engineering applications. 1976 edition. 560pp. 6 1/2 x 9 1/4. 0-486-46914-X

AN INTRODUCTION TO ORTHOGONAL POLYNOMIALS, Theodore S. Chihara. Concise introduction covers general elementary theory, including the representation theorem and distribution functions, continued fractions and chain sequences, the recurrence formula, special functions, and some specific systems. 1978 edition. 272pp. 5 3/8 x 8 1/2. 0-486-47929-3

ADVANCED MATHEMATICS FOR ENGINEERS AND SCIENTISTS, Paul DuChateau. This primary text and supplemental reference focuses on linear algebra, calculus, and ordinary differential equations. Additional topics include partial differential equations and approximation methods. Includes solved problems. 1992 edition. 400pp. 7 1/2 x 9 1/4. 0-486-47930-7

PARTIAL DIFFERENTIAL EQUATIONS FOR SCIENTISTS AND ENGINEERS, Stanley J. Farlow. Practical text shows how to formulate and solve partial differential equations. Coverage of diffusion-type problems, hyperbolic-type problems, elliptic-type problems, numerical and approximate methods. Solution guide available upon request. 1982 edition. 414pp. 6 1/8 x 9 1/4. 0-486-67620-X

VARIATIONAL PRINCIPLES AND FREE-BOUNDARY PROBLEMS, Avner Friedman. Advanced graduate-level text examines variational methods in partial differential equations and illustrates their applications to free-boundary problems. Features detailed statements of standard theory of elliptic and parabolic operators. 1982 edition. 720pp. 6 1/8 x 9 1/4. 0-486-47853-X

LINEAR ANALYSIS AND REPRESENTATION THEORY, Steven A. Gaal. Unified treatment covers topics from the theory of operators and operator algebras on Hilbert spaces; integration and representation theory for topological groups; and the theory of Lie algebras, Lie groups, and transform groups. 1973 edition. 704pp. 6 1/8 x 9 1/4. 0-486-47851-3

A SURVEY OF INDUSTRIAL MATHEMATICS, Charles R. MacCluer. Students learn how to solve problems they'll encounter in their professional lives with this concise single-volume treatment. It employs MATLAB and other strategies to explore typical industrial problems. 2000 edition. 384pp. 5 3/8 x 8 1/2. 0-486-47702-9

NUMBER SYSTEMS AND THE FOUNDATIONS OF ANALYSIS, Elliott Mendelson. Geared toward undergraduate and beginning graduate students, this study explores natural numbers, integers, rational numbers, real numbers, and complex numbers. Numerous exercises and appendixes supplement the text. 1973 edition. 368pp. 5 3/8 x 8 1/2. 0-486-45792-3

A FIRST LOOK AT NUMERICAL FUNCTIONAL ANALYSIS, W. W. Sawyer. Text by renowned educator shows how problems in numerical analysis lead to concepts of functional analysis. Topics include Banach and Hilbert spaces, contraction mappings, convergence, differentiation and integration, and Euclidean space. 1978 edition. 208pp. 5 3/8 x 8 1/2. 0-486-47882-3

FRACTALS, CHAOS, POWER LAWS: Minutes from an Infinite Paradise, Manfred Schroeder. A fascinating exploration of the connections between chaos theory, physics, biology, and mathematics, this book abounds in award-winning computer graphics, optical illusions, and games that clarify memorable insights into self-similarity. 1992 edition. 448pp. 6 1/8 x 9 1/4. 0-486-47204-3

SET THEORY AND THE CONTINUUM PROBLEM, Raymond M. Smullyan and Melvin Fitting. A lucid, elegant, and complete survey of set theory, this three-part treatment explores axiomatic set theory, the consistency of the continuum hypothesis, and forcing and independence results. 1996 edition. 336pp. 6 x 9. 0-486-47484-4

DYNAMICAL SYSTEMS, Shlomo Sternberg. A pioneer in the field of dynamical systems discusses one-dimensional dynamics, differential equations, random walks, iterated function systems, symbolic dynamics, and Markov chains. Supplementary materials include PowerPoint slides and MATLAB exercises. 2010 edition. 272pp. 6 1/8 x 9 1/4. 0-486-47705-3

ORDINARY DIFFERENTIAL EQUATIONS, Morris Tenenbaum and Harry Pollard. Skillfully organized introductory text examines origin of differential equations, then defines basic terms and outlines general solution of a differential equation. Explores integrating factors; dilution and accretion problems; Laplace Transforms; Newton's Interpolation Formulas, more. 818pp. 5 3/8 x 8 1/2. 0-486-64940-7

MATROID THEORY, D. J. A. Welsh. Text by a noted expert describes standard examples and investigation results, using elementary proofs to develop basic matroid properties before advancing to a more sophisticated treatment. Includes numerous exercises. 1976 edition. 448pp. 5 3/8 x 8 1/2. 0-486-47439-9

THE CONCEPT OF A RIEMANN SURFACE, Hermann Weyl. This classic on the general history of functions combines function theory and geometry, forming the basis of the modern approach to analysis, geometry, and topology. 1955 edition. 208pp. 5 3/8 x 8 1/2. 0-486-47004-0

THE LAPLACE TRANSFORM, David Vernon Widder. This volume focuses on the Laplace and Stieltjes transforms, offering a highly theoretical treatment. Topics include fundamental formulas, the moment problem, monotonic functions, and Tauberian theorems. 1941 edition. 416pp. 5 3/8 x 8 1/2. 0-486-47755-X

Mathematics–Algebra and Calculus

VECTOR CALCULUS, Peter Baxandall and Hans Liebeck. This introductory text offers a rigorous, comprehensive treatment. Classical theorems of vector calculus are amply illustrated with figures, worked examples, physical applications, and exercises with hints and answers. 1986 edition. 560pp. 5 3/8 x 8 1/2. 0-486-46620-5

ADVANCED CALCULUS: An Introduction to Classical Analysis, Louis Brand. A course in analysis that focuses on the functions of a real variable, this text introduces the basic concepts in their simplest setting and illustrates its teachings with numerous examples, theorems, and proofs. 1955 edition. 592pp. 5 3/8 x 8 1/2. 0-486-44548-8

ADVANCED CALCULUS, Avner Friedman. Intended for students who have already completed a one-year course in elementary calculus, this two-part treatment advances from functions of one variable to those of several variables. Solutions. 1971 edition. 432pp. 5 3/8 x 8 1/2. 0-486-45795-8

METHODS OF MATHEMATICS APPLIED TO CALCULUS, PROBABILITY, AND STATISTICS, Richard W. Hamming. This 4-part treatment begins with algebra and analytic geometry and proceeds to an exploration of the calculus of algebraic functions and transcendental functions and applications. 1985 edition. Includes 310 figures and 18 tables. 880pp. 6 1/2 x 9 1/4. 0-486-43945-3

BASIC ALGEBRA I: Second Edition, Nathan Jacobson. A classic text and standard reference for a generation, this volume covers all undergraduate algebra topics, including groups, rings, modules, Galois theory, polynomials, linear algebra, and associative algebra. 1985 edition. 528pp. 6 1/8 x 9 1/4. 0-486-47189-6

BASIC ALGEBRA II: Second Edition, Nathan Jacobson. This classic text and standard reference comprises all subjects of a first-year graduate-level course, including in-depth coverage of groups and polynomials and extensive use of categories and functors. 1989 edition. 704pp. 6 1/8 x 9 1/4. 0-486-47187-X

CALCULUS: An Intuitive and Physical Approach (Second Edition), Morris Kline. Application-oriented introduction relates the subject as closely as possible to science with explorations of the derivative; differentiation and integration of the powers of x; theorems on differentiation, antidifferentiation; the chain rule; trigonometric functions; more. Examples. 1967 edition. 960pp. 6 1/2 x 9 1/4. 0-486-40453-6

ABSTRACT ALGEBRA AND SOLUTION BY RADICALS, John E. Maxfield and Margaret W. Maxfield. Accessible advanced undergraduate-level text starts with groups, rings, fields, and polynomials and advances to Galois theory, radicals and roots of unity, and solution by radicals. Numerous examples, illustrations, exercises, appendixes. 1971 edition. 224pp. 6 1/8 x 9 1/4. 0-486-47723-1

AN INTRODUCTION TO THE THEORY OF LINEAR SPACES, Georgi E. Shilov. Translated by Richard A. Silverman. Introductory treatment offers a clear exposition of algebra, geometry, and analysis as parts of an integrated whole rather than separate subjects. Numerous examples illustrate many different fields, and problems include hints or answers. 1961 edition. 320pp. 5 3/8 x 8 1/2. 0-486-63070-6

LINEAR ALGEBRA, Georgi E. Shilov. Covers determinants, linear spaces, systems of linear equations, linear functions of a vector argument, coordinate transformations, the canonical form of the matrix of a linear operator, bilinear and quadratic forms, and more. 387pp. 5 3/8 x 8 1/2. 0-486-63518-X

Mathematics–Probability and Statistics

BASIC PROBABILITY THEORY, Robert B. Ash. This text emphasizes the probabilistic way of thinking, rather than measure-theoretic concepts. Geared toward advanced undergraduates and graduate students, it features solutions to some of the problems. 1970 edition. 352pp. 5 3/8 x 8 1/2. 0-486-46628-0

PRINCIPLES OF STATISTICS, M. G. Bulmer. Concise description of classical statistics, from basic dice probabilities to modern regression analysis. Equal stress on theory and applications. Moderate difficulty; only basic calculus required. Includes problems with answers. 252pp. 5 5/8 x 8 1/4. 0-486-63760-3

OUTLINE OF BASIC STATISTICS: Dictionary and Formulas, John E. Freund and Frank J. Williams. Handy guide includes a 70-page outline of essential statistical formulas covering grouped and ungrouped data, finite populations, probability, and more, plus over 1,000 clear, concise definitions of statistical terms. 1966 edition. 208pp. 5 3/8 x 8 1/2. 0-486-47769-X

GOOD THINKING: The Foundations of Probability and Its Applications, Irving J. Good. This in-depth treatment of probability theory by a famous British statistician explores Keynesian principles and surveys such topics as Bayesian rationality, corroboration, hypothesis testing, and mathematical tools for induction and simplicity. 1983 edition. 352pp. 5 3/8 x 8 1/2. 0-486-47438-0

INTRODUCTION TO PROBABILITY THEORY WITH CONTEMPORARY APPLICATIONS, Lester L. Helms. Extensive discussions and clear examples, written in plain language, expose students to the rules and methods of probability. Exercises foster problem-solving skills, and all problems feature step-by-step solutions. 1997 edition. 368pp. 6 1/2 x 9 1/4. 0-486-47418-6

CHANCE, LUCK, AND STATISTICS, Horace C. Levinson. In simple, non-technical language, this volume explores the fundamentals governing chance and applies them to sports, government, and business. "Clear and lively ... remarkably accurate." – *Scientific Monthly*. 384pp. 5 3/8 x 8 1/2. 0-486-41997-5

FIFTY CHALLENGING PROBLEMS IN PROBABILITY WITH SOLUTIONS, Frederick Mosteller. Remarkable puzzlers, graded in difficulty, illustrate elementary and advanced aspects of probability. These problems were selected for originality, general interest, or because they demonstrate valuable techniques. Also includes detailed solutions. 88pp. 5 3/8 x 8 1/2. 0-486-65355-2

EXPERIMENTAL STATISTICS, Mary Gibbons Natrella. A handbook for those seeking engineering information and quantitative data for designing, developing, constructing, and testing equipment. Covers the planning of experiments, the analyzing of extreme-value data; and more. 1966 edition. Index. Includes 52 figures and 76 tables. 560pp. 8 3/8 x 11. 0-486-43937-2

STOCHASTIC MODELING: Analysis and Simulation, Barry L. Nelson. Coherent introduction to techniques also offers a guide to the mathematical, numerical, and simulation tools of systems analysis. Includes formulation of models, analysis, and interpretation of results. 1995 edition. 336pp. 6 1/8 x 9 1/4. 0-486-47770-3

INTRODUCTION TO BIOSTATISTICS: Second Edition, Robert R. Sokal and F. James Rohlf. Suitable for undergraduates with a minimal background in mathematics, this introduction ranges from descriptive statistics to fundamental distributions and the testing of hypotheses. Includes numerous worked-out problems and examples. 1987 edition. 384pp. 6 1/8 x 9 1/4. 0-486-46961-1

Mathematics–Geometry and Topology

PROBLEMS AND SOLUTIONS IN EUCLIDEAN GEOMETRY, M. N. Aref and William Wernick. Based on classical principles, this book is intended for a second course in Euclidean geometry and can be used as a refresher. More than 200 problems include hints and solutions. 1968 edition. 272pp. 5 3/8 x 8 1/2. 0-486-47720-7

TOPOLOGY OF 3-MANIFOLDS AND RELATED TOPICS, Edited by M. K. Fort, Jr. With a New Introduction by Daniel Silver. Summaries and full reports from a 1961 conference discuss decompositions and subsets of 3-space; n-manifolds; knot theory; the Poincaré conjecture; and periodic maps and isotopies. Familiarity with algebraic topology required. 1962 edition. 272pp. 6 1/8 x 9 1/4. 0-486-47753-3

POINT SET TOPOLOGY, Steven A. Gaal. Suitable for a complete course in topology, this text also functions as a self-contained treatment for independent study. Additional enrichment materials make it equally valuable as a reference. 1964 edition. 336pp. 5 3/8 x 8 1/2. 0-486-47222-1

INVITATION TO GEOMETRY, Z. A. Melzak. Intended for students of many different backgrounds with only a modest knowledge of mathematics, this text features self-contained chapters that can be adapted to several types of geometry courses. 1983 edition. 240pp. 5 3/8 x 8 1/2. 0-486-46626-4

TOPOLOGY AND GEOMETRY FOR PHYSICISTS, Charles Nash and Siddhartha Sen. Written by physicists for physics students, this text assumes no detailed background in topology or geometry. Topics include differential forms, homotopy, homology, cohomology, fiber bundles, connection and covariant derivatives, and Morse theory. 1983 edition. 320pp. 5 3/8 x 8 1/2. 0-486-47852-1

BEYOND GEOMETRY: Classic Papers from Riemann to Einstein, Edited with an Introduction and Notes by Peter Pesic. This is the only English-language collection of these 8 accessible essays. They trace seminal ideas about the foundations of geometry that led to Einstein's general theory of relativity. 224pp. 6 1/8 x 9 1/4. 0-486-45350-2

GEOMETRY FROM EUCLID TO KNOTS, Saul Stahl. This text provides a historical perspective on plane geometry and covers non-neutral Euclidean geometry, circles and regular polygons, projective geometry, symmetries, inversions, informal topology, and more. Includes 1,000 practice problems. Solutions available. 2003 edition. 480pp. 6 1/8 x 9 1/4. 0-486-47459-3

TOPOLOGICAL VECTOR SPACES, DISTRIBUTIONS AND KERNELS, François Trèves. Extending beyond the boundaries of Hilbert and Banach space theory, this text focuses on key aspects of functional analysis, particularly in regard to solving partial differential equations. 1967 edition. 592pp. 5 3/8 x 8 1/2. 0-486-45352-9

INTRODUCTION TO PROJECTIVE GEOMETRY, C. R. Wylie, Jr. This introductory volume offers strong reinforcement for its teachings, with detailed examples and numerous theorems, proofs, and exercises, plus complete answers to all odd-numbered end-of-chapter problems. 1970 edition. 576pp. 6 1/8 x 9 1/4. 0-486-46895-X

FOUNDATIONS OF GEOMETRY, C. R. Wylie, Jr. Geared toward students preparing to teach high school mathematics, this text explores the principles of Euclidean and non-Euclidean geometry and covers both generalities and specifics of the axiomatic method. 1964 edition. 352pp. 6 x 9. 0-486-47214-0

Mathematics-History

THE WORKS OF ARCHIMEDES, Archimedes. Translated by Sir Thomas Heath. Complete works of ancient geometer feature such topics as the famous problems of the ratio of the areas of a cylinder and an inscribed sphere; the properties of conoids, spheroids, and spirals; more. 326pp. 5 3/8 x 8 1/2. 0-486-42084-1

THE HISTORICAL ROOTS OF ELEMENTARY MATHEMATICS, Lucas N. H. Bunt, Phillip S. Jones, and Jack D. Bedient. Exciting, hands-on approach to understanding fundamental underpinnings of modern arithmetic, algebra, geometry and number systems examines their origins in early Egyptian, Babylonian, and Greek sources. 336pp. 5 3/8 x 8 1/2. 0-486-25563-8

THE THIRTEEN BOOKS OF EUCLID'S ELEMENTS, Euclid. Contains complete English text of all 13 books of the Elements plus critical apparatus analyzing each definition, postulate, and proposition in great detail. Covers textual and linguistic matters; mathematical analyses of Euclid's ideas; classical, medieval, Renaissance and modern commentators; refutations, supports, extrapolations, reinterpretations and historical notes. 995 figures. Total of 1,425pp. All books 5 3/8 x 8 1/2.
Vol. I: 443pp. 0-486-60088-2
Vol. II: 464pp. 0-486-60089-0
Vol. III: 546pp. 0-486-60090-4

A HISTORY OF GREEK MATHEMATICS, Sir Thomas Heath. This authoritative two-volume set that covers the essentials of mathematics and features every landmark innovation and every important figure, including Euclid, Apollonius, and others. 5 3/8 x 8 1/2. Vol. I: 461pp. 0-486-24073-8
Vol. II: 597pp. 0-486-24074-6

A MANUAL OF GREEK MATHEMATICS, Sir Thomas L. Heath. This concise but thorough history encompasses the enduring contributions of the ancient Greek mathematicians whose works form the basis of most modern mathematics. Discusses Pythagorean arithmetic, Plato, Euclid, more. 1931 edition. 576pp. 5 3/8 x 8 1/2. 0-486-43231-9

CHINESE MATHEMATICS IN THE THIRTEENTH CENTURY, Ulrich Libbrecht. An exploration of the 13th-century mathematician Ch'in, this fascinating book combines what is known of the mathematician's life with a history of his only extant work, the Shu-shu chiu-chang. 1973 edition. 592pp. 5 3/8 x 8 1/2. 0-486-44619-0

PHILOSOPHY OF MATHEMATICS AND DEDUCTIVE STRUCTURE IN EUCLID'S ELEMENTS, Ian Mueller. This text provides an understanding of the classical Greek conception of mathematics as expressed in Euclid's Elements. It focuses on philosophical, foundational, and logical questions and features helpful appendixes. 400pp. 6 1/2 x 9 1/4. 0-486-45300-6

BEYOND GEOMETRY: Classic Papers from Riemann to Einstein, Edited with an Introduction and Notes by Peter Pesic. This is the only English-language collection of these 8 accessible essays. They trace seminal ideas about the foundations of geometry that led to Einstein's general theory of relativity. 224pp. 6 1/8 x 9 1/4. 0-486-45350-2

HISTORY OF MATHEMATICS, David E. Smith. Two-volume history – from Egyptian papyri and medieval maps to modern graphs and diagrams. Non-technical chronological survey with thousands of biographical notes, critical evaluations, and contemporary opinions on over 1,100 mathematicians. 5 3/8 x 8 1/2.
Vol. I: 618pp. 0-486-20429-4
Vol. II: 736pp. 0-486-20430-8